MATERIALS RESEARCH TO MEET 21ST-CENTURY DEFENSE NEEDS

Committee on Materials Research for Defense After Next
National Materials Advisory Board
Division on Engineering and Physical Sciences

NATIONAL RESEARCH COUNCIL
OF THE NATIONAL ACADEMIES

THE NATIONAL ACADEMIES PRESS
Washington, D.C.
www.nap.edu

The National Academies Press • 500 Fifth Street NW • Washington, DC 20001

NOTICE: The project that is the subject of this report was approved by the Governing Board of the National Research Council, whose members are drawn from the councils of the National Academy of Sciences, the National Academy of Engineering, and the Institute of Medicine. The members of the committee responsible for the report were chosen for their special competences and with regard for appropriate balance.

This project was conducted under Contract No. MDA972-01-D-001 from the U.S. Department of Defense. Any opinions, findings, conclusions, or recommendations expressed in this publication are those of the authors and do not necessarily reflect the views of the organizations or agencies that provided support for the project.

International Standard Book Number: 0-309-08700-7

Available in limited supply from:
National Materials Advisory Board
National Research Council
500 Fifth Street, N.W.
Washington, DC 20001
nmab@nas.edu

Printed in the United States of America

THE NATIONAL ACADEMIES

Advisers to the Nation on Science, Engineering, and Medicine

The **National Academy of Sciences** is a private, nonprofit, self-perpetuating society of distinguished scholars engaged in scientific and engineering research, dedicated to the furtherance of science and technology and to their use for the general welfare. Upon the authority of the charter granted to it by the Congress in 1863, the Academy has a mandate that requires it to advise the federal government on scientific and technical matters. Dr. Bruce M. Alberts is president of the National Academy of Sciences.

The **National Academy of Engineering** was established in 1964, under the charter of the National Academy of Sciences, as a parallel organization of outstanding engineers. It is autonomous in its administration and in the selection of its members, sharing with the National Academy of Sciences the responsibility for advising the federal government. The National Academy of Engineering also sponsors engineering programs aimed at meeting national needs, encourages education and research, and recognizes the superior achievements of engineers. Dr. Wm. A. Wulf is president of the National Academy of Engineering.

The **Institute of Medicine** was established in 1970 by the National Academy of Sciences to secure the services of eminent members of appropriate professions in the examination of policy matters pertaining to the health of the public. The Institute acts under the responsibility given to the National Academy of Sciences by its congressional charter to be an adviser to the federal government and, upon its own initiative, to identify issues of medical care, research, and education. Dr. Harvey V. Fineberg is president of the Institute of Medicine.

The **National Research Council** was organized by the National Academy of Sciences in 1916 to associate the broad community of science and technology with the Academy's purposes of furthering knowledge and advising the federal government. Functioning in accordance with general policies determined by the Academy, the Council has become the principal operating agency of both the National Academy of Sciences and the National Academy of Engineering in providing services to the government, the public, and the scientific and engineering communities. The Council is administered jointly by both Academies and the Institute of Medicine. Dr. Bruce M. Alberts and Dr. Wm. A. Wulf are chair and vice chair, respectively, of the National Research Council

www.national-academies.org

COMMITTEE ON MATERIALS RESEARCH FOR DEFENSE AFTER NEXT

HARVEY SCHADLER (*Chair*), General Electric Corporate Research and Development Center (retired), Schenectady, New York
ALAN LOVELACE (*Vice Chair*), General Dynamics Corporation (retired), La Jolla, California
JAMES BASKERVILLE, Bath Iron Works (General Dynamics), Bath, Maine
FEDERICO CAPASSO, Lucent Technologies, Murray Hill, New Jersey (until June 2000)
MILLARD FIREBAUGH, Electric Boat Corporation (General Dynamics), Groton, Connecticut
JOHN GASSNER, U.S. Army Natick Soldier Center, Natick, Massachusetts
MICHAEL JAFFE, New Jersey Center for Biomaterials and Medical Devices, Newark
FRANK KARASZ, University of Massachusetts, Amherst
HARRY A. LIPSITT, Wright State University (emeritus), Dayton, Ohio
MEYYA MEYYAPPAN, NASA Ames Research Center, Moffett Field, California
GEORGE PETERSON, U.S. Air Force Research Laboratory (retired), Wright-Patterson Air Force Base, Ohio
JULIA M. PHILLIPS, Sandia National Laboratories, Albuquerque, New Mexico
RICHARD TRESSLER, Pennsylvania State University (emeritus), University Park

Panel on Structural and Multifunctional Materials

HARRY A. LIPSITT (*Chair*), Wright State University (emeritus), Dayton, Ohio
MILLARD FIREBAUGH (*Vice Chair*), Electric Boat Corporation, Groton, Connecticut
MICHAEL I. BASKES, Los Alamos National Laboratory, Los Alamos, New Mexico
L. CATHERINE BRINSON, Northwestern University, Evanston, Illinois
THOMAS W. EAGAR, Massachusetts Institute of Technology, Cambridge
RICHARD J. FARRIS, University of Massachusetts, Amherst
D. DAVID NEWLIN, General Dynamics Land Systems, Sterling Heights, Michigan

GEORGE PETERSON, U.S. Air Force Research Laboratory (retired), Wright-Patterson Air Force Base, Ohio
RICHARD TRESSLER, Pennsylvania State University, University Park

Panel on Energy and Power Materials

JOHN GASSNER (*Co-chair*), U.S. Army Natick Soldier Center, Natick, Massachusetts
JAMES BASKERVILLE (*Co-chair*), Bath Iron Works, Bath, Maine
DANIEL H. DOUGHTY, Sandia National Laboratories, Albuquerque, New Mexico
SOSSINA M. HAILE, California Institute of Technology, Pasadena
ROBERT N. KATZ, Worcester Polytechnic Institute, Worcester, Massachusetts

Panel on Electronic and Photonic Materials

JULIA M. PHILLIPS (*Co-chair*), Sandia National Laboratories, Albuquerque, New Mexico
MEYYA MEYYAPPAN (*Co-chair*), NASA Ames Research Center, Moffett Field, California
HAROLD G. CRAIGHEAD, Cornell University, Ithaca, New York
NARSINGH B. SINGH, Northrop Grumman Corporation, Linthicum, Maryland
MING C. WU, University of California, Los Angeles
EDWARD ZELLERS, University of Michigan, Ann Arbor

Panel on Functional Organic and Hybrid Materials

FRANK KARASZ (*Chair*), University of Massachusetts, Amherst
LISA KLEIN, Rutgers University, Piscataway, New Jersey
VINCENT D. McGINNISS, Optimer Photonics, Columbus, Ohio
GARY E. WNEK, Virginia Commonwealth University, Richmond
LUPING YU, University of Chicago, Chicago, Illinois

Panel on Bioderived and Bioinspired Materials

MICHAEL JAFFE (*Chair*), New Jersey Center for Biomaterials and Medical Devices, Newark
ILHAN AKSAY, Princeton University, Princeton, New Jersey
MARK ALPER, University of California, Berkeley
PAUL CALVERT, University of Arizona, Tucson
MAURO FERRARI, Ohio State University, Columbus
ERIK VIIRRE, University of California, San Diego

National Materials Advisory Board Liaisons

ROBERT C. PFAHL, JR., Motorola (retired), Glen Ellyn, Illinois
KENNETH L. REIFSNIDER, Virginia Polytechnic Institute and State University, Blacksburg
EDGAR A. STARKE, University of Virginia, Charlottesville

Government Liaisons

ROBERT POHANKA, Office of Naval Research, Arlington, Virginia
ROBERT L. RAPSON, U.S. Air Force Research Laboratory, Wright-Patterson Air Force Base, Ohio
LEWIS SLOTER, Office of the Deputy Under Secretary of Defense (Science and Technology), Washington, D.C.
DENNIS J. VIECHNICKI, U.S. Army Research Laboratory, Aberdeen Proving Ground, Maryland
STEVEN WAX, Defense Advanced Research Projects Agency, Arlington, Virginia

National Materials Advisory Board Staff

ARUL MOZHI, Study Director
SHARON YEUNG DRESSEN, Program Officer (until July 2002)
JULIUS CHANG, Program Officer (until April 2002)
PAT WILLIAMS, Administrative Assistant

NATIONAL MATERIALS ADVISORY BOARD

JULIA M. PHILLIPS (*Chair*), Sandia National Laboratories, Albuquerque, New Mexico
JOHN ALLISON, Ford Research Laboratories, Dearborn, Michigan
FIONA DOYLE, University of California, Berkeley
THOMAS EAGAR, Massachusetts Institute of Technology, Cambridge
GARY FISCHMAN, Consultant, Palatine, Illinois
HAMISH L. FRASER, Ohio State University, Columbus
THOMAS S. HARTWICK, TRW (retired), Snohomish, Washington
ALLAN J. JACOBSON, University of Houston, Houston, Texas
SYLVIA M. JOHNSON, NASA Ames Research Center, Moffett Field, California
FRANK E. KARASZ, University of Massachusetts, Amherst
SHEILA F. KIA, General Motors, Warren, Michigan
ENRIQUE LAVERNIA, University of California, Davis
HARRY A. LIPSITT, Wright State University (emeritus), Dayton, Ohio
TERRY LOWE, Los Alamos National Laboratory, Los Alamos, New Mexico
ALAN G. MILLER, Boeing Commercial Airplane Group, Seattle, Washington
ROBERT C. PFAHL, JR., National Electronics Manufacturing Initiative, Herndon, Virginia
HENRY J. RACK, Clemson University, Clemson, South Carolina
KENNETH L. REIFSNIDER, Virginia Polytechnic Institute and State University, Blacksburg
T.S. SUDARSHAN, Materials Modification, Inc., Fairfax, Virginia
JULIA WEERTMAN, Northwestern University, Evanston, Illinois

National Materials Advisory Board Staff

TONI MARECHAUX, Director

Preface

The U.S. Department of Defense (DoD) requested that the National Research Council, through the National Materials Advisory Board (NMAB), conduct a study to identify and prioritize critical materials and processing research and development (R&D) that will be needed to meet 21st-century defense needs. The Committee on Materials Research for Defense After Next was established to investigate investments in R&D required to meet long-term (~2020) DoD needs. Its purpose was to explore revolutionary materials concepts that would provide an advantage to U.S. forces in weapons, logistics, deployment, and cost.

The committee was charged to address the following specific tasks:

- Review DoD planning documents and input from DoD systems development experts to identify long-term technical requirements for weapons system development and support.
- Develop materials needs and priorities based on DoD requirements.
- Establish and guide approximately five study panels to investigate identified priority areas and recommend specific research opportunities.
- Integrate and prioritize the research opportunities recommended by the study panels.
- Recommend ways to integrate materials and processes advances into new system designs.

The results of the initial phase, begun in December 1999, were documented in the January 2001 interim report.[1] In that initial phase the committee (13 scientists and engineers) met with technical representatives of the military services and DoD agencies, directors of service laboratories, and managers of DoD agencies (see Appendix A for a list of invited speakers). The objective of those meetings was to understand DoD's vision of current and future weapons, systems, and logistics requirements and its long-term cost targets. Although this aspect of the committee's study was not exhaustive, learning the status of current R&D supported by DoD, the U.S. Department of Energy, and the National Science Foundation provided a context for organizing subsequent meetings. The committee then met with materials experts from industry, academia, and national laboratories to identify research that could be brought to fruition in the 20- to 30-year time frame specified for the study. At a later meeting, the committee analyzed the data gathered and drafted the interim report.

In the next phase of the study, five technical panels were established (see Appendix B for the panel members' biographies):

- Structural and Multifunctional Materials,
- Energy and Power Materials,
- Electronic and Photonic Materials,
- Functional Organic and Hybrid Materials, and
- Bioderived and Bioinspired Materials.

These panels explored in depth the new opportunities in their areas of materials research and related them to DoD needs. Many of the concepts are still in their infancy. The questions the panels addressed were (1) What will be the impact of a successful materials R&D effort on future defense systems? and (2) How can the application of materials R&D be accelerated to meet DoD time constraints?

The organization of the panels by function encouraged technical experts to participate. Each panel was responsible for quantifying the impact of new materials and processes and for identifying technical roadblocks to their development. The technical panels were led by members of the study committee. NMAB liaisons to the study committee also served as

[1]National Research Council (NRC). 2001. *Materials Research to Meet 21st-Century Defense Needs—Interim Report.* Washington, DC: National Academy Press.

liaisons to the technical panels. This structure helped to ensure coherence of purpose, continuity of effort, and the rapid exchange of information.

We thank the committee and panel members for their participation in meetings and for their efforts and dedication in the preparation of this final report. We also thank the meeting speakers (listed in Appendix A) and participants and DoD study sponsors and liaisons, including Joseph Wells, Army Research Laboratory (retired), and Julie Christodoulou, Office of Naval Research. We thank the NMAB staff, especially Arul Mozhi, study director; Sharon Yeung Dressen, program officer; Julius Chang, program officer; Richard Chait, former staff director; Kevin Kyle, 2002 spring intern; Alan Lund, 2002 summer intern; Vikram Kaku, 2002 fall intern; and Pat Williams, administrative assistant.

This report has been reviewed by individuals chosen for their diverse perspectives and technical expertise, in accordance with procedures approved by the NRC's Report Review Committee. The purpose of this independent review is to provide candid and critical comments that will assist the institution in making its published report as sound as possible and to ensure that the report meets institutional standards for objectivity, evidence, and responsiveness to the study charge. The review comments and draft manuscript remain confidential to protect the integrity of the deliberative process. We wish to thank the following individuals for their participation in the review of this report:

Shaw Chen, University of Rochester;
David Clarke, University of California-Santa Barbara;
David Johnson, Jr., Agere Systems (retired);
David Kaplan, Tufts University;
James McBreen, Brookhaven National Laboratory;
Mark Reed, Yale University;
James Richardson, Potomac Institute for Policy Studies;
David Srolovitz, Princeton University;
Julia Weertman, Northwestern University;
Albert Westwood, Sandia National Laboratories (retired);
Mark Williams, National Energy Technology Laboratory; and
Yang Yang, University of California-Los Angeles.

Although the reviewers listed above have provided many constructive comments and suggestions, they were not asked to endorse the conclusions or recommendations, nor did they see the final draft of the report before its release. The review of this report was overseen by George Dieter, University of Maryland. Appointed by the National Research Council, he was responsible for making certain that an independent examination of this report was carried out in accordance with institutional procedures and that all review comments were carefully considered. Responsibility for the final content of this report rests entirely with the authoring committee and the institution.

Comments and suggestions can be sent via e-mail to NMAB@nas.edu or by fax to (202) 334-3718.

Harvey Schadler, *Chair*
Alan Lovelace, *Vice Chair*
Committee on Materials Research for Defense After Next

Contents

Figures and Tables

FIGURES

TABLES

Executive Summary

The U.S. Department of Defense (DoD) requested that the National Research Council (NRC), through the National Materials Advisory Board (NMAB), conduct a study to identify and prioritize critical needs for materials and processing research and development (R&D) to meet 21st-century defense needs. NMAB established the Committee on Materials Research for Defense After Next in the fall of 1999.

The committee identified DoD materials needs (described in Chapter 1) and explored the revolutionary defense capabilities that could result from R&D in five classes of materials:

- Structural and multifunctional materials,
- Energy and power materials,
- Electronic and photonic materials,
- Functional organic and hybrid materials, and
- Bioderived and bioinspired materials.

Due to the breadth of these materials areas, the committee established a separate panel to address each one; each panel produced a separate chapter (Chapters 3 through 7) that contains research priorities for its area (cross-referenced to other panel chapters as necessary). Each of these chapters begins with a summary that describes the panel's scope, DoD needs addressed, and R&D priorities identified. Chapter 8 integrates the R&D priorities from all five materials areas and presents the committee's R&D recommendations.

The committee recognized that realizing the revolutionary new defense capabilities that materials science and engineering offer will depend on more than just R&D; innovative management will also be needed to reduce risks in translating fundamental research into practical materials, and to promote cross-fertilization of scientific fields (e.g., biology and materials science) that heretofore have had little experience or contact with one another. Chapter 2 discusses these issues and presents the committee's recommendation for needed innovations in management.

This report describes the most promising areas for materials research and the systems that can benefit from them over the next 20 years. While these areas of research are expected to pay off handsomely for DoD, many of their benefits are likely to be evolutionary. However, the impact of the atomic bomb at the end of World War II reminds us that failure to invest in more speculative areas of research that involve extremely high risk but have comparably high potential payoff could lead to just the sort of technological surprises that DoD wishes to avoid. Therefore, the research opportunities identified in the present report should be addressed *in addition to* ensuring continued research at the forefront of physics, chemistry, biology, and materials science. Though this basic research may not result in technologies that are ready for deployment in 2020, it will provide the seeds for potentially revolutionary technologies to be realized later in the 21st century.

RECOMMENDATIONS

Management Processes

Discussions within the committee and its panels raised important questions about materials R&D management processes that affect all the broad areas of materials applications:

- How can emerging and future materials advances be better communicated to defense acquisition personnel who make technical decisions about the design of systems, subsystems, and components?
- How can technical decision makers in government and industry better communicate to materials researchers critical system, subsystem, and component engineering opportunities where existing materials limit system performance, so that the opportunities can be addressed by materials not yet in service?

• How can materials R&D be structured so that proper attention is paid to the complete set of functional characteristics a material must have to be put into service and perform successfully?

• How can materials R&D funds be deployed so as to reduce the time for a new material to progress from discovery or invention to service?

• How can government-funded materials R&D programs be prioritized to leverage commercial industrial materials R&D?

Addressing these questions can enhance the effectiveness of materials R&D in bringing promising materials from concept to service. To enable that effort, the committee offers the following recommendation.

RECOMMENDATION 1. TO ACCELERATE THE TRANSITION OF MATERIALS FROM CONCEPT TO SERVICE,

- **The Department of Defense (DoD) should budget research-to-development transition funds and devise a method to select early the materials advances on which to concentrate funds.**
- **DoD should adopt measures to enhance communication between materials researchers and users.**
- **DoD should make investments to organize and populate databases that describe the physical properties and attributes of materials to complement and validate materials computer modeling, and to facilitate communication among researchers and engineers at the system, subsystem, and component levels.**

Materials Research

The committee examined a broad range of materials research areas, from bioderived materials for wound healing to high-temperature structural materials for advanced jet engines and materials for advanced explosives and propellants. Despite this diversity, the committee was able to condense the results of its analysis into the following recommendations:

RECOMMENDATION 2. THE DEPARTMENT OF DEFENSE SHOULD MAKE RESEARCH INVESTMENTS IN THE DESIGN OF MATERIALS, DEVICES, AND SYSTEMS ASSISTED BY COMPUTATION AND PHENOMENOLOGICAL MODELS OF MATERIALS AND MATERIALS BEHAVIOR.

Early man developed and refined processes for melting and shaping metals and their alloys through exhaustive trial-and-error approaches extending over millennia, giving rise to the Bronze and Iron Ages. Such trial-and-error processes have continued into modern times, for example, the protracted search for an acceptable electric light bulb material, which culminated in the tungsten filament.

Today's breathtaking improvement in computational power enables materials scientists to move beyond trial and error and predict certain structures from first principles. Indeed, in some cases, such as energetic materials, computational approaches are leading experiments in new materials synthesis and are reducing the time required to discover and apply new materials. Used with phenomenological models to help identify and predict the characteristics and behavior of potentially revolutionary materials, advanced computational approaches offer a rapid and powerful means for discovery. The staggering potential benefits for DoD include:

- Improved ability to predict and select new materials,
- Design of materials with extreme properties,
- Improved design processes for structural composites,
- Protection of personnel and materiel against battlefield lasers,
- Improved materials for power generation, and
- Improved ability to predict and extend component life.

RECOMMENDATION 3. THE DEPARTMENT OF DEFENSE SHOULD MAKE RESEARCH INVESTMENTS THAT PROMOTE CONVERGENCE, COMBINATION, AND INTEGRATION OF BIOLOGICAL, ORGANIC, SEMICONDUCTOR, PHOTONIC, AND STRUCTURAL MATERIALS.

Convergence, *combination*, and *integration* are major themes for defense systems of the future. History has shown that major advances often occur at the points of convergence among disparate areas, whether in terms of broad fields of endeavor (e.g., chemistry, physics, and biology) or more specific areas (e.g., lithography and microsystems). These themes become apparent in the following R&D areas that the committee judged to be the most promising examples:

- Convergence between materials science and biology,
- New combinations of materials,

- Organic/inorganic composites,
- Integration of function in microsystems,
- Multifunctional materials,
- Preservation of biological function in devices,
- Organic materials for electronics and computing,
- Materials damage detection, and
- Materials processing and characterization.

RECOMMENDATION 4. THE DEPARTMENT OF DEFENSE SHOULD MAKE RESEARCH INVESTMENTS THAT PROMOTE DISCOVERY AND CHARACTERIZATION OF NEW MATERIALS WITH UNIQUE OR SUBSTANTIALLY IMPROVED PROPERTIES (BY 50 PERCENT OVER CURRENT PROPERTIES).

DoD systems of 2020 would benefit significantly from the discovery, development, and application of materials with properties that either are unique or considerably exceed those of today's materials. DoD weapon systems and platforms must be lethal and sustainable, and they must enhance the survivability of the user. These requirements drive new materials discovery.

Reducing volume and mass while enhancing functionality are key drivers for enhanced system performance. Munitions must be more compact; power and energy sources must have higher densities; armor must be lighter while providing equivalent or enhanced protection; platform structure should be lighter but remain strong in order to increase payload; and electronic and optical communication systems must be smaller and lighter while adding capability and bandwidth. All of these are tangible system requirements that depend on substantially improved new materials for use in subsystems, device components, and subcomponents. The most promising examples of such materials include:

- Tunable materials for infrared countermeasures,
- Improved organic photovoltaic materials,
- High-quantum-efficiency electroluminescent materials,
- Improved energetic materials,
- Improved optical materials,
- Agents that identify and interdict pathogens,
- Materials for efficient ultraviolet lasers and detector media, and
- Improved membranes.

RECOMMENDATION 5. THE DEPARTMENT OF DEFENSE SHOULD MAKE INVESTMENTS IN RESEARCH LEADING TO NEW STRATEGIES FOR THE PROCESSING, MANUFACTURE, INSPECTION, AND MAINTENANCE OF MATERIALS AND SYSTEMS.

Discovery of the structures and properties of novel materials and combinations of materials and integration of functionalities provide the seed for new and far-ranging DoD capabilities. This will likely require expert intuition, advanced computational methods, and combinatorial materials science. Miniaturized or highly parallel chemical reaction systems could be used to process novel materials for study or to create materials that might otherwise be difficult or impossible to process in bulk.

However, such an approach is only the starting point for introducing new materials into systems. Indeed, these materials must first be processed and demonstrated successfully in the laboratory on a small scale before there can be practical and robust methods for scale-up to sufficient quantities. Moreover, techniques must be available for fabricating and assembling new materials into prescribed geometries, for quality assurance to make certain they meet requirements, and for inspecting them in the field to identify service-induced defects and other problems. It is also crucial that there be means for repairing them, and for assuring the adequacy of repair. Examples of the most promising areas for investment to meet these requirements include:

- Materials characterization while in service,
- Self-repairing or self-healing materials,
- Processing that yields material with high purity,
- Large-scale processing of nanomaterials, and
- Biomimetic and bioinspired materials manufacturing.

CONCLUSIONS

Future defense systems could employ advanced materials that are self-healing, can interact independently with the local environment, and can monitor the health of a structure or component during operation. Advanced materials could act as a host for evolving technologies, such as embedded sensors and integrated antennas. Advanced materials must also deliver traditional high performance in structures; protect against corrosion, fouling, erosion, and fire; control fractures; and serve as fuels, lubricants, and hydraulic fluids.

The next 20 years will present the materials community with daunting challenges and opportunities. Requirements for material producibility, low cost, and ready availability will be much more demanding than they are today. On the other hand, spurred by the accelerated pace of advances in electronics and computation, the performance, life span, and maintainability of materials will be greatly enhanced. Some of the advances will result from R&D undertaken by commercial enterprises for competitive advantage in areas like telecommunications and computation. In other areas, however, DoD may have to bear the funding burden directly. In these special areas, considerable funding will be necessary not only to identify critical new materials, but also to accelerate their progress through development to applications in the defense systems of the future.

CHAPTER ONE

Department of Defense Materials Needs

GENERIC DEFENSE NEEDS

As the United States, its institutions, and its citizens interact throughout the world, situations may arise that call for military force. To safeguard its interests for the foreseeable future the United States must be able to project military power around the globe.[1] Today, the United States is far and away the greatest military power in the world and is far ahead in using new technologies in military systems.

Because oceans surround the United States, it has in place a worldwide base structure to support forward-deployed forces. Whereas other nations tend to operate from their own territory, as a matter of strategic principle the United States projects military power over long distances with medium-range and short-range systems. The oceans form a buffer over which the United States maintains military control. The buffer is not impermeable, though. It can be penetrated by long-range missiles, space-based systems, and submarines.

The present capability of the United States to project military power around the globe came about mainly as a result of U.S. participation in World War II, which left an infrastructure of U.S. military bases around the globe to support treaty obliga-

[1]Most of the information in this section is based on presentations to the committee by Andrew Marshall, director, Office of Net Assessment for the Secretary of Defense: Marshall, A., "Overview of DoD Vision and System Needs," paper presented to the Committee on Materials Research for Defense After Next, National Research Council, Washington, DC, February 15, 2000; Marshall, A., "Overview of DoD Vision and System Needs—An Update," paper presented to the Committee on Materials Research for Defense After Next, National Research Council, Washington, DC, January 29, 2002.

tions and mutual-defense agreements. In the aftermath of the Cold War, the infrastructure of overseas bases has been largely dismantled, partly to reduce costs and partly because many host countries no longer accept a powerful U.S. presence on their soil. The closing of overseas bases affects all of the Armed Forces, but especially the Army and the Air Force. The Navy, in effect, brings its overseas bases with it in the form of the fleet, taking advantage of freedom of the seas to move about the world.

Low-cost, highly capable commercial technologies are increasingly enabling many nations, including some with very limited resources, to mount regional threats based on precision strikes from their own territories. It is recognized that in the next 20 to 30 years, the period of interest for this study, even ships of the U.S. fleet might be threatened in the home waters of most nations, though submarines can be expected to remain relatively invulnerable. The possibility that even resource-constrained nations will be able to acquire potent offensive capabilities—though of limited range—increases the vulnerability of overseas U.S. bases.

The following core tasks therefore lie ahead for the U.S. military:

- Projecting long-distance military power;
- Maintaining capability to fight far away;
- Coping with the eroding overseas base structure;
- Safeguarding the homeland; and
- Adjusting to major changes in warfare, including joint-service operations, coalition peacekeeping, and an increased number of humanitarian missions.

The following trends in warfare are expected to continue:

- The need will increase for a precision strike force that can maneuver rapidly and effectively and can survive an attack while far away.
- The force must be able to conceal its activities from an enemy while detecting enemy activities.
- Advances in information technology will increase coordination among forces. Global awareness through real-time networked sensors and communications will facilitate command and control and enable precision strikes.
- Using unmanned vehicles, information will be gathered in new ways, military power will be delivered remotely, and the risk of casualties will be reduced.

- Fighting in urban areas will increase, requiring entirely different strategies and equipment.
- Guerilla warfare will require new strategies and weapons.

Marshall and other speakers also focused the committee's attention on new threats that could not be counteracted by force projection.[2] Weapons of mass destruction, for example, are a growing threat. During the Cold War, when nuclear weapons, principally in the Soviet Union, were a major concern, U.S. security was safeguarded by strategic deterrence to neutralize that threat. Among threats to the United States may be the delivery by missile or other means of small numbers of nuclear, chemical, or biological weapons from very disparate sources, including terrorist groups, or an assault on the complex web of information systems that are becoming increasingly important in the delivery of goods and services. Vulnerable infrastructure points include power grids and dams. The terrorist acts of September 11, 2001, highlight the need to address such threats.[3]

Marshall and others urged that the United States maintain its capability to project military power over long distances, harness advancing technologies to maintain its technological lead as long as possible (recognizing that other nations will be working to counter U.S. capabilities), continue to control the ocean buffer, and make effective plans for safeguarding the homeland.

EXAMPLES OF SYSTEM NEEDS

Briefings by senior officials of the Armed Services, defense agencies, and other government agencies covered a variety of short-term and long-term perspectives. The starting point for the fundamental needs of the U.S. Army, Michael Andrews, deputy assistant secretary of the Army for research and technology, said, was that though it would continue to be based in the United States the army would have to be able to respond

[2]Vickers, M., "The Revolution in Military Affairs (RMA)," paper presented to the Committee on Materials Research for Defense After Next, National Research Council, Washington, DC, February 15, 2000; Henley, L., "The Revolution in Military Affairs After Next," paper presented to the Committee on Materials Research for Defense After Next, National Research Council, Washington, DC, February 15, 2000.

[3]National Research Council (NRC). 2002. *Making the Nation Safer: The Role of Science and Technology in Countering Terrorism.* Washington, DC: National Academies Press.

quickly to provide a global presence.[4] Quick response would necessarily be provided by airlift, which implies lightweight forces. The Army has a goal of being able to move a large concentration of troops anywhere in the world in 48 hours. Because it anticipates that armored and airborne forces will continue to be essential for attacking an enemy, the Army will emphasize highly mobile lightweight vehicles equipped with next-generation armor and stealth to survive against high-intensity threats. Infantrymen face increasingly potent weapons and require a very high degree of information connectivity on the battlefield. Currently, each soldier must carry heavy personal equipment and batteries. In the future, the Army plans to reduce each soldier's load by using lighter-weight equipment, low-power electronics, and robotics offloading.

The Air Force also envisions supporting its military power projection from the United States.[5] According to Kenneth Harwell, chief scientist for the Air Force Research Laboratory, the goal is to deliver munitions from the United States to targets anywhere around the globe in less than an hour.[6] This will require both very high speeds and very lightweight material. Meeting this goal carries a formidable technical challenge. The Air Force also envisions the need for increased emphasis on space assets.

To fulfill its objective of decisively influencing events on land anywhere at any time, the Navy wants systems that are stealthy and can operate in littoral areas around the world.[7] The Navy is emphasizing antisubmarine and mine warfare to ensure that the U.S. fleet can carry out its mission in inshore waters.

The goal of the Marine Corps is to provide very lightweight, agile, early-entry forces, operating from sea bases with minimal needs for logistic

[4]Andrews, M., "Army Vision and S&T: Accelerating the Pace of Transformation," paper presented to the Committee on Materials Research for Defense After Next, National Research Council, Washington, DC, February 15, 2000.

[5]Delaney, L., "Air Force Modernization," paper presented to the Committee on Materials Research for Defense After Next, National Research Council, Washington, DC, February 15, 2000.

[6]Harwell, K., "Air Force Research Laboratory: Technology Vision," paper presented to the Committee on Materials Research for Defense After Next, National Research Council, Washington, DC, February 16, 2000.

[7]DeMarco, R., "Department of the Navy Science and Technology—Materials: Today, Tomorrow, and the Future," paper presented to the Committee on Materials Research for Defense After Next, National Research Council, Washington, DC, February 15, 2000.

support ashore.[8] The Marine Corps will be a strike force, not an occupying force.

All the military services expressed a need for systems that cost less and require less maintenance. The principal delivery hardware—ships, submarines, aircraft, and vehicles—will probably be expected to remain in service for very long periods, placing new demands on underlying technologies for durability, maintainability, and ease of upgrade. The Air Force was particularly forceful in stating the case for aircraft with very long service lives and the need to maintain and modernize aircraft at much lower cost.

TRANSLATION TO MATERIALS AND PROCESS NEEDS

Though presentations to the committee were organized by the needs of individual services, materials needs are related to more generic systems, platforms, and equipment. For example, because all the services require aircraft, materials research that leads to more advanced aircraft will be valuable to all the services. Although the need of a particular service might be the impetus for meeting a defined capability, once a technology matures to the point that it can be readily used in an operational system, it may also be used advantageously in similar systems for other purposes.

Ships, submarines, aircraft, military vehicles, sailors, airmen, soldiers, and marines of the future will all need advanced materials that enable significant changes in maneuverability (mobility, speed, agility); force protection (from nuclear, biological, chemical, kinetic, or explosive weapons through stealth, identification, armor, and active defense); engagement (highly concentrated and sustained firepower); and logistics (durability, maintainability).

Advanced materials must satisfy diverse requirements for speed, strength, precision, survivability, signature, materials selection, cost, weight, and commonality. Ships may be able to travel at speeds in excess of 75 knots; very lightweight tanks will travel at speeds up to 75 miles per hour; weapons will be delivered at hypersonic speeds. Materials will have

[8]Gray, A., "Thoughts on Future Marine Corps Materials Needs," paper presented to the Committee on Materials Research for Defense After Next, National Research Council, Washington, DC, February 15, 2000.

to endure tougher environments for longer periods—from ocean depths to Arctic cold to desert heat to space reentry. For example, the Army envisions new high-strength, very lightweight materials that can be integrated with primary structures and can defend ground vehicles against future weapons. The Army seeks to field systems with an offensive capability similar to the M1A2 Abrams tank at about one-third the weight; these may include unmanned ground vehicles and munitions.

Materials processes will require greater precision compatible with reduced fabrication and operational tolerances. Increased survivability will require materials that can reduce multispectral signatures (e.g., radio frequency, thermal, acoustic) and provide ballistic protection. The cost of acquisition and lifetime support of DoD platforms and war fighters must be reduced. For example, precision munitions will not be completely effective until they are inexpensive enough to be used by even the lowest tactical unit. Materials that increase capability but also increase cost must be compared to materials that provide current capabilities at reduced costs, including maintenance and upkeep costs.

Because manpower is the single largest DoD cost, materials that reduce the need for manpower will be extremely beneficial. Materials that reduce weight but retain functionality will permit increases in payload and range. It will be necessary to use common materials across platforms, between services, and among soldiers, sailors, airmen, and marines. The services can all benefit from processes that encourage sharing of materials technologies.

Desired Materials Properties

DoD needs various types of functionality, alone and in combination, for military systems. This section describes types of materials, combination of materials properties, and engineering issues that new materials must address in defense systems.

R&D in materials and processes will be required to improve existing materials and achieve breakthroughs in new materials and combinations. Examples of the types of materials needed are as follows:

- *Lightweight materials that provide equivalent functionality.* A pervasive requirement for DoD systems is weight reduction, in everything from tanks to the equipment carried by each soldier. Yet, military forces will require at least the same functionality as today's systems—for example, the ability to withstand enemy fire—as well as new functionalities—particularly in sensing surroundings, communicating with and responding to other elements of the force, increasing lethality, and responding to new threats.

- *Materials that enhance protection and survivability.* It will remain critical for DoD forces to be able to withstand enemy fire. Research must be done on new approaches to providing this capability, particularly at reduced weight. Materials that perform multiple functions are a promising area of research.
- *Stealth materials.* As the range of operations increases, it will be increasingly important for some elements to remain invisible for as long as possible using stealth technologies. This is an important area of research, particularly in multifunctional structural materials that incorporate a stealth capability and electronic and/or optical materials and devices that may actively respond to probes to achieve invisibility.
- *Electronic and photonic materials for high-speed communications.* Communication and coordination of elements in tomorrow's force will require extremely broad bandwidth and secure transmission, reception, and interpretation. These, in turn, will require materials that will enable these functions, be they optical, electronic, or some combination thereof. In addition, the force elements will be under constraints on weight, speed, or both—meaning that the desired functionality must be achieved in very small volume with as little weight as possible.
- *Sensor and actuator materials.* A battlefield of interconnected elements poses many demands, one of which is the need to detect signals that may be of many types. Research on sensor and actuator materials and their integration into larger systems will be critical to DoD. An emerging area of need is detection of chemical and biological agents.
- *High-energy-density materials.* There is a need for explosive materials that have higher energy per unit mass than current explosives, the properties of which can be tailored to maximize lethality against specific threats. Similarly, successful R&D on new high-energy-density propellants could make it possible to decrease the mass and increase the range of projectiles.
- *Materials that improve propulsion technology.* Improved propulsion will be necessary on land and sea, in air and space. The numerous materials issues this raises range from the need for high-energy-density

fuels to materials for improved undersea propulsion. Materials for hypersonic propulsion systems are a priority for future Air Force systems.

Materials Characteristics

In addition to the materials needs already discussed, a number of other materials properties are desirable for DoD systems. These characteristics cut across all the classes of materials that have been mentioned. Although they would not of themselves lead to the selection of a particular material for a given application, they are likely to affect that selection. Examples of these characteristics are as follows:

- *Multifunctionality.* One way to reduce weight and volume is by using materials that can perform at least two functions (e.g., stealth and structural support). Multifunctionality can be thought of on two scales: (1) mesoscopic (e.g., coatings) or macroscopic (e.g., load-bearing), and (2) microscopic or nano, in which multiple physical phenomena are produced through molecular design or architectural texture. The concept of multifunctionality encompasses many classes of materials and applications: Structural materials may be self-interrogating or self-healing, provide stealth, or protect against enemy fire; microscopic materials or systems may combine sensing, moving, analyzing, communicating, and acting.
- *Self-healing and self-diagnosing.* Materials with self-healing and self-diagnosing characteristics address a number of DoD needs, from improving survivability to minimizing system maintenance. The advantages and concomitant savings are obvious.
- *Low total system cost.* Cost is a primary criterion for DoD decisions for the foreseeable future. Materials that result in low total system cost (including initial, operating, maintenance, and disposal costs) will have a decided advantage, even if their cost is high relative to other choices.
- *Low maintenance.* Because systems for Defense After Next are likely to be in use for many decades, materials that do not require extensive, active maintenance are clearly preferred. An example is the Navy's critical need to reduce the necessity of paint chipping, a resource-intensive task that adds significantly to the Navy's manpower requirements.
- *High reliability.* The consequences of materials failure in DoD systems can be dire. The need for materials and manufacturing processes that are highly reliable spans all classes of materials.

- *Environmental acceptability.* DoD has become increasingly sensitive to the environmental impact of military activities. The use of environmentally friendly energy sources and the efficient use of power are areas where R&D is needed. Materials and processes for DoD systems should have as little adverse impact on the environment as possible. An area of special concern is the decommissioning and disposal of obsolete systems, including the recycling and reuse of as much of an old system as possible.

Engineering Issues

Successful research on broad classes of materials and processes will affect future defense systems only if other engineering issues are considered. The issues discussed below must be considered concurrently with R&D because they are likely to reveal the directions that such research should take.

Design Methods

The successful introduction of a new material into a system requires that the material chosen be integrated with the design of the entire system. Conversely, early in the design process, the design can be modified to compensate for the shortcomings of a material. Comprehensive databases of material properties would greatly facilitate the design of materials. Characterizing materials by desired properties, an essential aspect of materials design, may require new techniques.

Materials by Design

If system designers and materials and process experts communicate early in the process, it may be possible to design or tailor a material to meet the needs of the system, rather than designing the system around available materials. In fact, this is likely to become more common as strategic experiments, coupled with validated physical models, are incorporated into computer simulations that enable the virtual exploration of composition, structure, processing, and properties as a partial replacement for extensive laboratory experimentation. A move in this direction would also reduce development time and cost.

The properties of a given material are seldom ideal for all aspects of a particular application. Materials and process scientists must be aware of the desirability of materials or combinations of materials with properties that can be tailored to be compatible with a variety of system requirements. The concept of treating multiple materials as a system (e.g., in structural composites) has already been shown to increase the tailorability of materials substantially. The ability to tailor materials in multiple dimensions or more quickly than has previously been possible should also open new doors.

Influence of Materials on Development and Deployment Costs

The choice of a material can affect development and deployment costs in many ways. First, the cost of raw materials may be an issue, particularly if it is very high. If a material is extremely difficult to process or if there is a great deal of waste or rejected material, costs can also escalate quickly. Creating an entirely new processing technology or qualification procedure is also a cost factor. It is critical that the baseline process, the range of acceptable properties, and other factors governing these costs be understood early on so that the research program can perhaps tailor the research to respond to them.

Availability of Commercial Alternatives

Buying a material or a part will almost always be cheaper than designing, building, and manufacturing a new one specifically for DoD. In some cases, the quality of the commercial product may even be better, thanks to economies of scale and quality controls. Before DoD invests in developing a new material or process to improve performance, the improvement should be shown to have a potential payoff high enough to be worth the investment. A crucial aspect of DoD's investments in the development and acquisition of the best materials in the future will be choosing between in-house development, collaboration with industry, or purchase.

Risk Management

For obvious reasons, DoD is highly risk averse. In the materials and process arena, risk aversion translates to a reluctance to introduce new materials or processes unless the benefits have been clearly demonstrated and the risk is acceptably low. Minimizing risk implies minimizing the number of materials in use and identifying materials that simplify component or system design, which could reduce cost and risk—assuming, of course, that the material meets all of the other risk minimization criteria. Finally, DoD must constantly be on the lookout for fatal flaws that could eliminate a material from consideration no matter how desirable its other characteristics might be.

Manufacturing

A new material or process that seems very promising in the laboratory may be useless because it is not manufacturable. Materials and process scientists must ensure that production of the material or component can be scaled to a level appropriate for its end use. Defect density must be acceptably low and the yield high so that there is little or no waste or inefficiency in the process. It must be possible to inspect and characterize the product, either through rigorous process-based quality approaches or standard inspection. Finally, the product must be manufacturable at an acceptable cost.

Life-Cycle Issues

How a material will perform over the life of the component or system is an essential consideration in choosing a material for a particular application. If there is a predictive reliability model for a material or process, the expected lifetime for the material in use can be easily determined. An alternative would be a self-interrogating, self-reporting material or system that indicates when attention is required. Materials incorporated into DoD systems must be highly reliable. This can be ensured in a number of ways, ranging from selecting materials with important properties that persist over a wide range of conditions to rigorous process-based quality that allows for confident prediction of reliability. The components and systems must be maintainable so that they can function at specification for many years, if necessary. All life-cycle costs, including the cost of maintenance, must be considered. Finally, what will happen at the end of system life must be considered, taking into account recycling or reuse of as much of the system as possible and environmentally conscious disposal of the rest.

CHAPTER TWO

Using New Materials in Defense Systems

INTRODUCTION

Beyond the technical requirements for future military materials, the committee and its panels raised important questions about the research and development (R&D) management processes that broadly affect materials application:

- How can emerging and future materials advances be better communicated to defense acquisition personnel who make technical decisions about the design of systems, subsystems, and components?
- How can technical decisionmakers in government and industry better communicate to materials researchers critical system, subsystem, and component engineering opportunities where existing materials limit system performance, so that the opportunities can be addressed by materials not yet in service?
- How can materials R&D be structured so that proper attention is paid to the complete set of functional characteristics a material must have to be put into service and perform successfully?
- How can materials R&D funds be deployed so as to reduce the time for a new material to progress from discovery or invention to service?
- How can government-funded materials R&D programs be prioritized to leverage commercial industrial materials R&D?

Similar questions were addressed and recommendations made in *Materials Science and Engineering: Forging Stronger Links to Users* (NRC, 1999). The same questions were addressed in a different context in *Materials Technologies for the Process Industries of the Future: Management Strategies and Research Opportunities* (NRC, 2000). The process industries report differentiates between a "technology push" and a "market pull" research strategy, noting that the Department of Energy, Office of Industrial Technology, was transitioning its research programs to the latter in the interest of getting more materials research to produce results that would quickly be applied in industry.

EXPEDITING THE USE OF NEW MATERIALS IN DEFENSE SYSTEMS

Concurring with the previous studies, the committee concluded that the time required to move new materials from research to service generally seems excessive. That the same questions continue to arise in both military and commercial contexts suggests that the problem is fundamental. Much of the problem appears to lie in the difficulty of communicating needs and potential solutions among system, subsystem, and component engineers and materials scientists and engineers.

To expedite the use of new materials in defense systems, the committee identified three important and interconnected ideas, discussed below.

Research-to-Development Transition Funding

Funds need to be allocated specifically to take promising materials from research through the developmental steps necessary for the materials to become reasonable candidates for service use. Advanced materials that are highly leveraged because they can be used widely or because they have the potential to significantly enhance performance in an important defense capability need rapid development focused on engineering issues like manufacturability, low cost, inspectability, joinability, and environmental impact.

The materials research elements of defense agencies and departments meet annually to coordinate their efforts. The annual meeting gives them the opportunity to identify high-leverage materials early in the research and provide funding to expedite their investigation. Analysis of materials research that has transitioned effectively, and research that has not, could produce useful criteria for prioritizing projects.

Some corporations invest substantially in their own research-to-development efforts, using analytical methods to make the business case for key lines of research that could be applied by DoD. The efficiency of transition will be positively influenced by high-level participation of system designers and engineers. The objective of research-to-development transition funding is to reduce the risk of putting materials into service by making the new material attractive to the engineering and programmatic staff responsible for the performance of deployable systems.

Communications

Communications among materials researchers and system, subsystem, and component engineers are required to get materials into use. These communities already cooperate in many ways, but it is obvious that their communications are often inadequate. Most materials choices are made by working engineers, generally in industry, responsible for performing on a contract. Direct connections are needed to relate materials R&D with industry system, subsystem, and component engineers. The formal process for translating research into systems ends up in many interpretations and reinterpretations of needs, with much information lost and time wasted.

Acquisition reformers have mandated the elimination of military specifications (MILSPECS) and standards. This process was a traditional vehicle for deploying proven technology created through service-funded R&D. Underlying the decision to move away from MILSPECS was the assumption that commercial influences would be more effective in facilitating incorporation of advanced technology (NRC, 2002), but their elimination has removed one mechanism for incorporating technology. Though few would want MILSPECS back, new methods for better connecting the value-added links in the chain from basic research through deployment would be useful. Another important aspect of the problem of communication between materials researchers and practicing engineers is that in very new fields, there are no established communication links. For example, there may be biological or bioinspired materials that can complement or supplant traditional materials in areas where the biologists have no experience or contacts and the engineers have no knowledge base to draw their attention to these researchers.

Databases

More attention needs to be paid to the many ways in which data about materials are organized and made available to engineers in industry who specify material for systems, subsystems, and components. High-quality, validated, physical data captured in databases are essential to:

- Support reliable, physically verified computer modeling;
- Communicate the readiness for service of specific materials with defined performance criteria;
- Identify functional gaps in the R&D for given materials;
- Identify materials that are being explored by government and industry;
- Provide engineers with management options for resolving system performance and cost issues; and
- Categorize traditional and emerging materials, their functions, and their multifunctional capabilities in a standardized taxonomy.

Data integrity is of paramount importance. Databases need to be free of organizational or commercial bias. Because the database is a vehicle for communications, the kinds of data entered, the formats, quality standards for data acceptability, and other aspects of the database must be decided not only by materials scientists and engineers but also by system, subsystem, and components engineers—all parties to the communication must be represented. The committee envisions that DoD materials research staffs can give structure to the process in which materials data are validated, presented, formatted, and reviewed by letting DoD's interest be known and stimulating progress through grants and contracts.

CONCLUSIONS

This committee believes that many of the materials derived from current and future research will be of great value in engineered systems to meet pressing future defense needs. The effectiveness of materials research and development in bringing advanced materials from concept to service can be enhanced by addressing questions and ideas discussed. Two recent NRC reports (1999 and 2000) and this report confirm the importance of doing so. Implementing the following three-part recommendation will accelerate the transition of materials from concept to service.

RECOMMENDATION 1. **TO ACCELERATE THE TRANSITION OF MATERIALS FROM CONCEPT TO SERVICE,**

- **The Department of Defense (DoD) should budget research-to-development transition funds and devise a method to select early the materials advances on which to concentrate funds.**
- **DoD should adopt measures to enhance communication between materials researchers and users.**
- **DoD should make investments to organize and populate databases that describe the physical properties and attributes of materials to complement and validate materials computer modeling, and to facilitate communication among researchers and engineers at the system, subsystem, and component levels.**

REFERENCES

National Research Council (NRC). 1999. Materials Science and Engineering: Forging Stronger Links to Users. Washington, DC: National Academy Press.

National Research Council (NRC). 2000. Materials Technologies for the Process Industries of the Future: Management Strategies and Research Opportunities. Washington, DC: National Academy Press.

National Research Council (NRC). 2002. The Impact of Acquisition Reform on Department of Defense Specification and Standards for Materials and Processes. Washington, DC: National Academy Press.

CHAPTER THREE

Structural and Multifunctional Materials

CHAPTER SUMMARY

The Panel on Structural and Multifunctional Materials focused on emerging materials and the processes used for their fabrication, with special attention to the types of multifunctionality that could be designed into a material. An example might be a composite material in which both the matrix and filaments serve several functions: The matrix might contain microcapsules sensitive to mechanical stress that, upon breaking, would highlight the damaged area by changing color. The strengthening filaments might have two different compositions which, when imbedded in a conducting polymer matrix, would produce a galvanic current. Such a material might be the basis for a new generation of lightweight, long-service electric vehicles.

This chapter discusses DoD structural materials development approaches and goals. It highlights the importance of lighter, stiffer, and stronger materials, and the need for materials to operate for long periods at high temperature with predictable degradation. These materials are necessary to improve vehicle mobility, maneuverability, transportability, and survivability.

Once all the data were presented, the panel identified four areas of R&D opportunity. In priority order these are

- Materials design assisted by computation,
- Service-induced material changes,
- Composite materials design and development, and
- Integration of nondestructive inspection and evaluation into the original design.

These four opportunities are expanded upon, with special emphasis on the design of structural materials that are truly multifunctional.

Investments in these research areas should result in advances that would yield many of the necessary new DoD materials. Such advances will

- Reduce development time and costs,
- Modernize design criteria,
- Predict and verify functionality,
- Continuously monitor in-service health, and
- Predict residual life.

INTRODUCTION

The Panel on Structural and Multifunctional Materials looked at (1) emerging materials and processes for fabricating structural materials and (2) multifunctionality that could be built into the structure, such as health monitoring, thermal-load dissipation, and electromagnetic radiation management. This panel concentrated on mesoscopic and macroscopic multifunctionality scales, such as thin laminates, mesoscopic trusses, active fibers, and coatings. The Functional Organic Panel (see Chapter 6) addressed microscopic multifunctionality introduced by atomic or molecular design. Using some of the polymer matrices discussed by that panel can open a fruitful area for future investigation: composite materials in which several levels of multifunctionality are incorporated with the structure.

Using data presented by many outstanding scientists and engineers, the panel identified four broad R&D opportunities; in order of priority they are

- Materials design assisted by computation,
- Service-induced material changes,
- Composite materials design and development, and
- Integration of nondestructive inspection and evaluation into original design.

These opportunities, if exploited, can produce many DoD materials of the future. They should also reduce the time and resources required for development of new materials, change the design criteria, and insure that functionality is predicted and measured and that behavior of the materials in use is continuously evaluated.

The concept of multifunctionality spans all four opportunities. There is a need to maximize the potential for use of materials with intrinsic multifunctionality and to design and fabricate composites with active multifunctional phases.

This panel believes that computational materials science is the most important opportunity in the future of materials research. Thus, "materials design assisted by computation" is a major opportunity to

- Design better materials,
- Better understand their behavior,
- Design better structures with them, and
- Shorten the development cycle from concept to implementation.

As a result of the improved understanding emerging from computation-assisted design of materials, it will also become possible to design better structures using these materials. Thus, the second opportunity from this panel is entitled "service-induced material changes" or "how to better use materials." The rationale for this revolutionary concept is the thought that materials and structures ought to be designed for functionality rather than being based on the material's initial properties. To design this way it is necessary to understand history-dependent properties and performance evolution. This means the science base for the use of materials must be expanded, and nonequilibrium structures and materials must be understood in order to support a science-based constitutive theory that is explicit in the extensive variables of the material. The advantages of such an approach would be to reduce materials and system development cost by virtual engineering and simulation during design and speed up materials and systems development, enabling higher performance and longer-life designs.

Given the timeframe for this study (~2020), the panel did not address possible incremental improvements of materials, unless they could lead to a breakthrough in a system capability. Rather, the focus was on materials in which one might expect major (20-25 percent) performance improvements over the next 15 to 25 years. As Table 3-1 shows, monolithic materials are unlikely to show performance gains of this magnitude; composite materials—or at least materials combinations—are much more likely to do so.

TABLE 3-1 Potential for Achieving Property Improvements of 20 to 25 Percent over Current State of the Art for Various Classes of Materials by 2020

	Strength	Toughness	Stiffness	Density	Environmental Resistance	High Temperature Capability
Metals	No	No	No	No	Yes	Increase by 200°F
MMCs[a]	Yes	No	Yes	Yes	Yes	No
Ceramics	Yes	No	No	No	No	No
CMCs[a]	Yes	Yes	Yes	Yes	No	Yes
Polymers	Yes	Yes	Yes	No	No	No
PMCs[a]	Yes	Yes	Yes	Yes	Yes	Yes

[a]MMC = metal matrix composite; CMC = ceramic matrix composite; PMC = polymer matrix composite. Processing of these materials is critical to improvements in the properties indicated.

Since the fractions of fibers comprising a composite material may be varied over a wide range, these materials may be designed with a broad range of density, stiffness, and strength values, because the filaments used are very strong and have large elastic moduli. Many filamentary materials also have lower density than metals, so metal matrix composites, for example, can be less dense than the parent metal but have greater strength and stiffness, increasing both specific strength and specific stiffness. These points are especially relevant for polymer matrix composites where very high specific properties can be obtained for nominal temperature applications. Thus, "composite materials design and development" is a compelling approach that merits study and refinement, because reduced weight is a primary design criterion in many structures.

In addition to extensive composite material development efforts, this panel believes that opportunities exist for "integrating non-destructive inspection and evaluation into the original design" of both materials and structures. This would allow for continuous monitoring of the health of all newly designed structures. Integrating sensors into the structure requires that they be very small, so many new types of sensors must be created. In addition, small portable advanced sources, such as X-ray and neutron sources, will be needed to allow field evaluation of structures and some sources should be incorporated into the internal structure in places that would be difficult to examine with an outside source.

This chapter continues with a discussion of DoD needs for materials that are lighter, stiffer, and stronger than those available today, and materials capable of long-term high-temperature exposure without excessive degradation. These materials are needed to improve vehicle mobility, maneuverability, transportability, and survivability. The following sections describe applications of computational approaches to materials design, the integration and optimization of materials systems, multifunctional materials, materials with self-healing abilities, materials with thermal or electrical conductivity spanning the range from conductors to insulators, and advanced coatings and adhesives. The chapter concludes by drawing out crosscutting materials R&D opportunities that will be vital to critical DoD systems in the coming decades.

DOD NEEDS FOR MULTIFUNCTIONAL STRUCTURAL MATERIALS

The materials needs of DoD are all-inclusive. Though DoD uses every known type of structural material, it has a continuing need for new materials because the military must always strive to be better prepared than any potential enemy, and potential enemies are themselves also striving to become better prepared.

In general, the military needs materials that are lighter, stronger, stiffer, and usable at higher temperatures. This allows equipment to be more mobile, maneuverable, transportable, and to last longer. For the military's air arms, the goals have always been to fly higher, farther, and faster. Recently, DoD has emphasized the total life-cycle cost of all types of equipment and materials of construction. There has been great interest in "smart materials"—e.g., materials that will monitor and report on their own health. This requires the development of many new sensors, some of which must be an integral part of the material. New instruments to activate and query these sensors are also required. Finally, there have recently been demands for multifunctional materials, e.g., a composite with high strength and stiffness in which the strengthening filaments can supply battery power.

The significance of materials costs to integrated systems costs is discussed in Appendix C, but the essential points are covered here. Forty years ago, Westbrook[1] noted that structural materials vary in price by seven orders of magnitude and that the usage of a material in pounds per annum is inversely related to the cost per pound; for example, reducing

[1]Westbrook, J.H., Internal General Electric Report, General Electric, Schenectady, NY, 1962.

TABLE 3-2 Market for Structural Materials

	Value of Pound of Weight Saved Over Life of Vehicle ($)	Number of Units Sold Per Year	Market Size ($ billion)
Automobiles and trucks	2	30 million	600
Commercial aircraft	200	2,500	150
Spacecraft	20,000	100	20

materials cost by a factor of two can result in a four-fold increase in usage. For a vehicle, another important factor is the value of weight saved over its service life. Table 3-2 summarizes the impact on the market when gasoline is $2.00 per gallon, an automobile has a 100,000-mile life, commercial aircraft has a 100,000-hour life, and spacecraft goes into orbit once.

The cost of materials is a relatively small fraction of the fabricated cost of a structure, typically 10 to 20 percent. Thus, combining these fabricated costs of the structure with the value of a pound saved gives the average maximum cost that can be tolerated in a particular application. For the automobile example, where the value of a pound saved is $2.00 times a 20 percent material cost as a fraction of total cost, the upper limit is $0.40 per pound for the primary structural material of the automobile, which is about the cost of automobile-quality steel. It is also possible to conclude that aluminum will not be a cost-effective substitute for steel in automobiles until gasoline costs $4.00 per gallon.[2] As noted in Appendix C, these calculations must be fine-tuned to align with factors such as the speed at which the object moves and the complexity of the structure fabricated. For example, the value of a pound saved in the rotating part of the gas turbine in an airplane is 10 times the value of a pound saved in the fuselage. Also, materials costs for complex composite structures are as little as 2-5 percent of the total fabricated cost.

The importance of reducing materials weight is that all vehicles, engines, and aircraft can be lighter and require less energy to run, thus saving fuel or enabling them to carry larger payloads. For military vehicles, fuel savings are especially critical. Figures from the U.S. Army indicate that

[2]P. Bridenbaugh, Alcoa (retired), private communication, October 11, 2001.

70 percent by weight of the expendable supplies needed on a battlefield are fuels. The cost of a gallon of fuel, transported halfway around the world, stored and transported to forward fuel depots, and finally delivered by air to vehicles at the battlefront may be as much as $400 per gallon, although the more usual price was between $13 and $30 per gallon (DSB, 2001). Weight reduction coupled with strength increase is doubly important for vehicle or engine weight and for allowable engine size. For rotating or oscillating components, both mass and strength are important, but the impact of reducing rotating or oscillating mass cannot be minimized. In any engine, reduction of moving mass makes it possible to reduce the mass of shafts, bearings, and bearing support structures; thus a simple weight reduction in moving mass can cascade through the engine to a dramatic total weight decrease. These same factors apply to flywheel-type energy storage systems and to rail-gun systems.

Material stiffness is a property that is especially important for extensive structures that must hold their shape and in tube and sheet structures where buckling propensity is directly related to the elastic modulus (Seely and Smith, 1955). Specific modulus, a performance index of a material, is defined as the elastic modulus divided by density. In many sheet or column structures where the material has a low modulus, to ensure sufficient component rigidity the thickness of the structure must be increased above that required to achieve the specified design strength. In the design of these structures, the most important quantity is the elastic modulus. The structural stresses may not be the significant quantities—that is, the stresses may not limit the loads that can be applied to the member without causing structural damage, and hence the strength properties of the material (such as yield stress) are not of primary importance. Note that the specific moduli of Fe, Al, and Mg are almost identical, so simply substituting one material for another does not change this (ASM, 1961).

One area where elastic modulus is critical is in the design and construction of gun tubes. High stiffness allows the propellant charge to expand and accelerate a projectile to maximum velocity, but does not allow the expanding gases to leak past the projectile and thus undermine efficiency. New high-energy propellants require that gun tubes be stiffer; this can be achieved using a composite reinforced by filamentary windings. These propellants also demand that the refractory interior surfaces of the gun tube be resistant to erosion and corrosion at temperatures higher than those that are currently encountered.

In monolithic materials, stiffness is very difficult to change by alloying and is hardly affected by microstructure. In fact, alloying and heat treatment work better to decrease stiffness than to increase it. Thus, designers of new monolithic materials have only two options: The material must be made lighter so that structural efficiency is improved or it must be made stronger. These, then, are the most promising directions for monolithic materials research.

In the design of composite materials, however, density, stiffness, and strength are almost independent quantities (Schaffer et al., 1995). The filaments used are quite strong and often have elastic moduli about twice the modulus of metals, so increasing stiffness is a real design opportunity. Many filamentary materials also have low density; incorporating them into a metal or alloy will both decrease density and increase strength, thus increasing both the specific strength and the specific stiffness. It is these qualities that make new composite materials compelling opportunities for the future. Because ceramic or polymer matrix materials can produce very strong, lightweight structures, research on these materials is of primary importance.

The drawbacks are that the fabrication costs of composite materials often exceed those of metals and alloys, and their ductility and fracture toughness are usually lower. Where cost is not of primary importance and low ductility and global toughness are not the primary causes of failure, composite materials are prime candidates for structural design. However, R&D on composite materials must proceed in many directions simultaneously. The most important needed advances are in the processing and scale-up of multiple material combinations. Current processing methods are slow, involving hand lay-up of binder-impregnated composite sheets or the processing of only small batches of material. Other needed advances include adequate exterior coatings for nonoxide composites, fiber coatings to prevent fiber-matrix chemical diffusion at high temperatures, and control of the interface between filaments and matrix to optimize properties.

It is also essential to understand the mechanics of fracture in each type of material. Especially in the design and manufacture of composite materials, there are opportunities for integrating nondestructive investigation and evaluation sensors into the original design of both materials and structures. This would allow for continuous health monitoring of all newly designed structures. The integration of sensors into the structure requires that they be very small, so many new types of sensors will be needed.

The integration of sensors into the surface of DoD land vehicles is a critical step in the formulation of smart armor (TACOM, 2001). The surface sensor must pick up the speed and direction of the penetrating projectile and initiate the appropriate response in the armor package to defeat the threat. Shape charge jets travel the fastest and require a defeat mechanism different from the long rod penetrator. The material at the base of the armor to capture the resultant debris needs to be lightweight but strong enough to prevent perforation. The defeat mechanism of smart armor requires the storage and distribution of power, as discussed in Chapter 4.

Much effort has been expended in recent years on materials for propulsion and power generation systems, because these offer the greatest potential benefits. For example, the operating efficiency of a gas-turbine engine will increase by more than 1 percent for every 10 degrees centigrade increase in the turbine-inlet gas temperature (Sims, Stoloff, and Hagel, 1987). The increase in efficiency may be either increased power or decreased fuel usage. Substantial fuel savings in aircraft and ground vehicle engines and in stationary turbines used for local power generation can be achieved by using new materials that can accommodate the temperature increase.

All devices that operate at high temperature demand special materials that respond to an extensive set of design requirements. Strength and stiffness are just two of these. Others include creep and fatigue resistance coupled with good oxidation and corrosion properties. In-service conditions provide additional challenges, including ingestion of debris. Thus, the ability to absorb damage without compromising safety is a crucial characteristic. Because a common measure of aircraft turbine engine efficiency, especially in military systems, is the thrust-to-weight ratio, R&D strategies often use material elements to increase thrust (through higher-temperature alloys) and decrease engine weight (with low-density materials).

Higher turbine-inlet gas temperatures have become possible for three fundamental reasons. First, single crystal turbine blades were introduced into the early turbine stages. These blades can operate at higher temperatures because elements added to strengthen and stabilize grain boundaries lowered the alloy melting temperature. More recently, new higher temperature alloys contain increasing amounts of Ta and Re, both heavy elements, so the density of these blade materials has increased by about 25 percent (Antolovich, 1992; Kissinger, 1996; Pollock, 2000). Finally, thermal barrier coatings are used to insulate metallic material from the heat of flowing gases.

Improvements in the insulating characteristics of the thermal barrier coatings should be pursued. However, the other techniques mentioned are approaching the limit of their usefulness. Adding dense elements to the blade is self-defeating because the heavier rotating blades increase stresses on the supporting disk. Thus, the disk, the shaft, and the bearing support structure must be strengthened, adding still more weight to the engine. Significant progress in turbine engine efficiency is likely to come from materials that allow engine weight to decrease.

New high-temperature materials are also needed for nuclear reactors. The hottest materials in a gas fluid reactor may be as high as 1000°C. Solid oxide fuel cells are also typically operated at about 1000°C, a temperature at which fuel cell reactions occur efficiently, although lower temperature operation (to ~600°C) is being pursued to reduce material and manufacturing costs. For military purposes, where the input is presumed to be diesel fuel, cells operating at the higher temperatures will still be required. In these systems there is also a great need for coatings for refractory metals and for technology to join refractory metals to baser metals. In some applications, such as propulsion for ships, materials must perform at these very high temperatures for very long periods without maintenance (for example, modern naval nuclear plants operate for their nominal 30-year life without refueling).

Other areas where marked progress is possible are metallic and ceramic-based composites (NRC, 1998), perhaps using discontinuous reinforcements and amorphous and nanoscale matrix materials. Current efforts could lead to the emergence of such materials. These efforts should be guided by computational design approaches, and should also consider material degradation in service as part of alloy selection and treatment. Finally, since periodic inspection of parts already in an assembled engine is not easy, the new materials should have NDI/NDE elements integrated into their structure. New high-temperature materials are needed for continuous exposure to 1400°C. These materials must balance the properties needed for use in complex systems. In the current environment the necessary bywords, in addition to high temperature, stiffness, and strength, are affordability, durability, and reliability.

Reasonable goals for materials development by 2020 would be a 25 percent increase in tensile, creep, and fatigue strength for a whole range of materials, coupled with a 25 percent decrease in density. Of necessity there can be no degradation in oxidation resistance at temperatures to about 1000°C. Materials usable to about 1200-1400°C with the same density as superalloys (not more than ~10 g/cc) will also be required. High modulus materials are also needed for space frames, vehicle body stock, tubular construction materials, etc. By 2020 many materials with modulus increases of 20-40 percent will be required. It is likely that these modulus goals can only be attained with composites.

SPECIFIC AREAS OF OPPORTUNITY

Materials Design Assisted by Computation

A critical limitation to the introduction of new materials into military systems is the extreme length of the development cycle, often as long as 15 years. At the front end of the development cycle is the arduous task of creating a material with superior properties. Not only must the composition of the material be established, but—of equal importance—so must the process by which this material is made. Processing of structural materials is critical because the microstructure developed during processing is a primary determinant of the resultant mechanical properties. Complicating the design process is the fact that a material has many mechanical and physical properties, e.g., elastic modulus, yield strength, ductility, fracture toughness, fatigue resistance, density, weldability, and corrosion resistance. Improvement of one property often causes degradation of another. Hence material design is a compromise. Traditionally, the development cycle is long because it involves sequential synthesis of the material and then extensive testing of many properties and combinations.

Computation holds the promise of shortening the development cycle by eliminating much experimental synthesis and testing. However, to replace experiment, computation must be reliable; the computational tools must consistently predict a suite of mechanical, physical, and thermodynamic properties of materials. Specifically, not only must the properties of single crystal multicomponent materials be predictable; the properties of multiphase, nonequilibrium materials as a function of processing conditions must also be predictable. Not only is this task not easy, at this time it is not possible. Some current limitations of computational materials science at the atomistic scale are

- Lack of understanding of the underlying atomic mechanisms that determine phase transformations and mechanical properties, e.g., strength, durability, and damage tolerance;

- Inability to query long enough times for full development of microstructures and deformation at realistic strain rates;
- Inability to process large enough samples to give meaningful microstructures; and
- Inadequate knowledge of potential interactions in multicomponent systems.

Currently it is possible to model certain materials processes (e.g., casting, forming, and heat treatment) using commercially available finite element method (FEM) codes. These simulations have proved extremely valuable for process design. Empirical material constitutive models, obtained from experiment, are a key input into FEM simulations. Unfortunately, because these models are material-specific (only one composition) and can rarely predict microstructures, they are essentially useless for materials design. Development of computational materials science would enable mechanistic constitutive materials models to replace the empirical models now used in FEM simulations. The result will be a predictive tool that would allow materials designers to shorten the experimental synthesis and testing stages, and hence development cycle time and cost, by at least one order of magnitude.

A key question is how to develop these computational tools. The properties of materials are determined by the quantum mechanical interaction of electrons. Currently it is possible to model these interactions quite well for a system of hundreds of atoms at a temperature of absolute zero. Advances in computational speed over the next two decades should make it possible to expand this capability considerably. This panel believes that the way to extend electronic calculations into a useful constitutive model is through a series of models at various size scales. A schematic of the process is shown in Figure 3-1 as the materials model. The immediate goal is to be able to perform predictive calculations at the field level, e.g., FEM calculations, using a series of calculations at various length and time scales.

Each model will use input from the next smaller scale. This approach depends on having models for every microstructural level from electrons to atoms to dislocations to grains to macroscopic materials. For any given problem, the approach would be to begin by determining what can be learned at the electron level. That information is then distilled and used to formulate an approach at the atomic level; new things learned there are then used to formulate a computational approach at the next level, and so

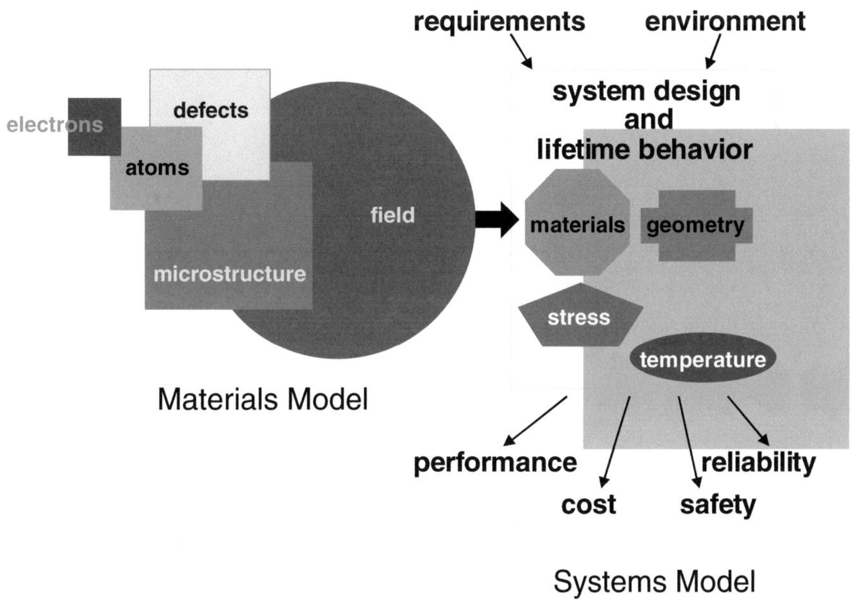

FIGURE 3-1

Schematic of materials and systems interactions through a series of models at various size scales.

forth. The minimum amount of information that captures the phenomena at each level is passed to the next level. At every level of calculation, experimental information is added both in model development and in verification. Thus, by sequential computations, where each level of computation is rationalized against existing knowledge, a picture of a complex phenomenon can be constructed.

The panel identified three key issues for the future:

1. How to balance the tradeoff between reliability and computational speed for interatomic potentials for various material classes and multicomponent systems;
2. How to extract information at each level (i.e., how to connect to larger size and longer time models and visualization of mechanisms); and
3. How to ensure that the appropriate mix of scientists and engineers is available to develop the required methods and perform the necessary calculations.

The panel then assessed the difficulty of the tasks in the path to robust computational materials design over the next two decades. In order of increasing difficulty, they are:

1. Prediction of equilibrium phase stability in multicomponent materials;
2. Prediction of thermomechanical processing (nonequilibrium phase transformations and microstructure evolution); and
3. Prediction of mechanical behavior like deformation and fracture.

To be a useful tool in designing materials, the computational process must succeed in at least the first two predictions. Scientifically based predictive materials constitutive models can then be developed that would be critical enablers of the prediction of life-cycle behavior, the topic of the next section.

Model-Based Life-Cycle Sensing and Prediction

The challenge of bringing advances in science to bear on societal engineering needs requires robust models that interlace science, technology, economics, manufacturing, and management. Given the speed of technological change and the pressing need for rapid deployment of technology for military needs, the excessive time and cost involved in the manufacture and testing of numerous candidate materials constitute a barrier that must be overcome. The most promising approach is to use science to create robust models for design and development in virtual environments.

Such an approach is illustrated in the right-hand portion of Figure 3-1. The key to lifetime design is a sophisticated systems model that can predict system performance and reliability throughout service lifetime; system costs, including initial, maintenance, and disposal costs; and system safety both in normal operation and under threat conditions. This panel's view of the future has a systems engineer optimizing all these requirements, not only by changing conventional design parameters (e.g., geometry) but also selecting materials (both composition and processing) through the materials model. The materials model is a critical enabling component of the systems model.

Perhaps the greatest cost of not addressing this challenge is in the areas of reliability, safety, and serviceability. All private firms that make human service devices (from cell phones to airplanes) must design their products to global standards, with a reliability of from 3σ to 5σ with a confidence rating typically of 90-95 percent. Traditional methods of determining reliability take two general approaches. The older approach uses factors of safety to knock down measured values to design levels to allow a margin of safety (which is actually a margin of ignorance and cost). Factors of safety are still widely used in the heavy and high-performance equipment industries—exactly those that influence military deployment costs. The second approach is to design on the basis of statistical analysis of data. If data are sufficient, it is possible to estimate allowable design levels with the required confidence. Today, however, neither of these approaches is adequate to the available science and pressing economic and military requirements. The issues this situation brings up include the following:

1. Factors of safety are not "safe"; they are not based on an understanding of what makes something safe. They only help us to avoid failure because of ignorance. Consequently, we always pay the price of ignorance (usually by overdesign) when factors of safety are used. We cannot continue to pay that price given current economic constraints.
2. Statistical designs are inherently flawed when applied to life prediction or reliability estimates, for two fundamental reasons. First, statistical data, which are generally collected at the beginning of life, do not represent the nature of the behavior near the end of life. Changes in properties and performance of modern materials (such as tires, turbine parts, and composite components) over their useful life may be significant. Second, no basic physics, chemistry, mechanics, economics, or other precepts associated with property or performance changes during life are incorporated into statistical designs. They are based on only one embodiment of all the variables.

This study has identified a need to put a science base under degradation and property/performance evolution so that mechanistic models can be used for life-cycle design. The vision is to have a sufficiently fundamental understanding of phenomena like corrosion, creep, stress and creep rupture, galvanic action, radiation effects, and long-term temperature effects on a wide range of engineering materials to enable the development of in situ and other real-time property sensors and mechanistic science-based performance models. This will allow us to predict the life-cycle behavior of materials and components using such sensor information in real time.

The objective should be to establish a modeling and computational philosophy that supports the following benefits of this technology:

1. History-based reliability prediction for *individual* components in airplanes, vehicles, ships, etc. to greatly reduce the cost of maintenance and greatly increase the safety and reliability of engineering devices.
2. Robust, multidisciplinary virtual design environments that greatly reduce the cost of design and testing, prototyping, manufacturing, and deployment and increase the speed of development of engineering components.
3. Performance simulations that support tradeoff studies before costs are incurred for tooling and manufacturing.
4. Industrial management models that support decision making at all stages of engineering design and development.

A central theme of this work should be the integration of materials science, manufacturing science and technology, and management science and technology. The expected benefit of this effort, based on current efforts to approach this problem in isolated instances in the aircraft industry, includes a reduction of cost of development by one third and a decrease in the time of development by about 50 percent.

Multifunctional Materials

Multifunctionality in the context of materials applications is often referred to as "structure +," because one of the functions of the material is typically structural—to carry load or define shape. The other functions can be as varied as actuation (to control position, shape, or load), electrical (to provide insulating or conducting pathways), thermal (for heat transfer control), health (for monitoring deformation or damage), stealth (for managing electromagnetic or visible signature), or self-healing (for embedded mechanisms for repairing localized damage).

As with the other major sections of this report, the concept of multifunctional materials applies to all the major classifications of materials (polymers, metals, ceramics), but by its very nature it is most prevalent in composites of these materials. Any material classified as "smart" or "intelligent" also is multifunctional, usually due to microstructural/molecular level response characteristics that manifest as macroscopic actuation ability with electric, magnetic, or thermal input. Table 3-3 shows possible materials phenomena that could have military applications. Table 3-4 shows examples of military applications. An important global issue related to multifunctionality is therefore the understanding and modeling of materials in a hierarchical fashion to address key length scales, an issue discussed in detail above. The following subsections will briefly discuss the future of multifunctionality, mentioning first the general area of smart materials, then the possibilities in composites.

Smart Materials

Smart or intelligent materials where revolutionary improvements in material design or application are likely are piezoelectric materials, shape memory alloys (SMAs), and active polymers. Piezoelectric materials are electroelastic materials that change their length in certain directions when an electric field is applied or that develop an electric field upon mechanical loading due to dipole moments within the crystalline lattice structure. The panel believes that traditional polycrystalline piezoelectric materials will see only evolutionary changes in the next 20 years as improved design methods enhance actuation strain, but that single-crystal piezoelectrics offer a greater possibility for substantial gains due to their inherently larger actuation strains.

SMAs, of which NiTi is the most common, can provide large stress or strain response to temperature actuation due to a diffusionless solid-state-phase transformation from a low-temperature martensitic phase to a high-temperature austenitic phase. Although SMAs have been used since the 1960s, thorough understanding of the microstructural mechanisms and response to nonuniaxial, nonisothermal loading is lacking. Current efforts aimed at addressing these deficiencies will yield significant advances in shape memory materials. The greatest strides in the next 20 years are anticipated in magnetic, thin-film, and porous microstructure SMAs and in SMA composites.

For some applications one drawback of SMAs in general has been sluggish high-frequency response due to the times required for heating and cooling the material. This problem can be resolved by magnetic SMAs that are activated through a magnetic field instead of temperature and by thin-film SMAs

TABLE 3-3 Examples of Multifunctional Capabilities of Targeted Structural Materials

	Smart Material			Composite			
Functionality	Piezoelectric Material	Shape Memory Alloy	Polymer	Foam	Polymer Matrix Composite	Metal Matrix Composite	Ceramic Matrix Composite
Health monitoring		Sensor element	Coated damage sensors		Optical fibers		
Self-healing		As composite or foam	Crack healing	SMA foam	Via polymer chemistry	SMA 2nd phase	
Electromagnetic and acoustic wave management				Acoustic	Layered and reinforced		Erosion resistance
Actuation	Single crystals	Films, magnetic	Active polymer valves		SMA/piezoelectric fibers layers	SMA 2nd phase	
Thermal				Insulation	Thermal management	Discontinuously reinforced	Infrared management
Electrical					Nanoreinforced systems, circuit boards	Discontinuously reinforced	
Magnetic		Magnetic activated			Data storage	Permanent magnetic	Permanent magnetic

TABLE 3-4 Examples of Military Applications Likely to Benefit from Revolutionary Advances in Multifunctional Structural Materials

	Application						
Functionality	Armor	Stealth	Antennas and Arrays	Engines	Structure	Power	Launchers
Health monitoring	Remote sensing	Remote sensing	Remote sensing	Remote sensing	Remote sensing	Remote sensing	Remote sensing
Self-healing	Shape memory alloys and foams	Stealth coating	Conformal and directional	Corrosion resistant			Corrosion resistant
Electromagnetic and acoustic wave management		Embedded nanoscale networks for shielding					
Actuation	Smart armor		Embedded conformal antennas				
Thermal		Infrared matching		Higher operating temperature		Higher operating temperature	Higher operating temperature
Electrical	Smart armor	Millimeter wave reduction			Photovoltaics	Power storage	
Magnetic						Permanent magnetics	

because of their high surface-to-volume ratio. Further understanding of fabrication and response of thin-film SMAs will make it possible to use them in microelectromechanical systems (MEMS) and even nanoelectromechanical systems (NEMS) devices. The materials are currently very experimental due to problems of maintaining uniformity in fabrication, understanding of fatigue mechanisms, and lack of constitutive descriptions at small scales. Exploration of shape memory effects, which have been demonstrated for some polymeric solids, is a promising avenue to be explored.

Porous SMA materials offer the advantages of other metallic foam structures enhanced by material response due to the activity of the SMA matrix material. Porous SMAs have been fabricated successfully under less than ideal conditions, resulting in promising structures of high porosity, though with extremely uneven properties. With emphasis on processing methods, microstructure control and optimization, and modeling to address an active material, SMA foams offer great potential for lightweight, damage-tolerant, self-healing structural components.

The last category of SMAs with potential to emerge as a vital material system in the future is composites, where the SMA could be a secondary phase to provide self-healing. The self-healing aspect is discussed more fully in the next section.

Electroactive polymers and ion-exchange polymer-metal composites are active material systems that show large deformations in the presence of a low applied voltage, mimicking biological tissue. The polymers have conductivities in a wide range, from values comparable to semiconductors to values as high as copper. Their conductivity and color can be reversed by controlled chemical or electrical stimuli. Electroactive polymers can be used both as light and chemical sensors and as actuators, but because of their low elastic modulus they are not typically used as structural members. Nevertheless, they are certainly multifunctional materials with enormous growth potential and could play a role in self-healing composite materials.

Composites

Composites offer the greatest opportunity for significant advances in material design and function. Because of their multiphase nature and fabrication methods these materials offer simple routes for embedded sensors, actuators, and other elements that provide multifunctionality. In the traditional ceramic matrix composite (CMC), metal matrix composite (MMC), and polymer matrix composite (PMC) materials, evolutionary developments in embedded optical sensors for local strain interrogation are likely, as are small improvements in the use of embedded smart materials (SMAs or piezoelectric elements) as actuators. Revolutionary advances in traditional composites should result from use of nanotechnology and wireless technology and incorporation of self-healing mechanisms.

Nanotechnology offers exciting opportunities for radical changes in composite functionality. One area is to incorporate carbon nanotubes or other nanoparticulates into the matrix material to reach the percolation threshold at relatively low volumes (e.g., 1 percent or less). If the orientation and patterning of the nanoscale reinforcement could be controlled, integrated, and organized, the networks could add functions to health monitoring, surveillance, and stealth. A high impact of nanoscale reinforcements is foreseen, however, only in PMCs, where they also offer potential for improved mechanical properties. In particular, nanoscale reinforcement of the polymer could enhance matrix properties to the point where compression strength is improved; entangled networks of nanotubes may also toughen the material.

Understanding the mechanisms and the degree of property improvement possible needs significant research attention. In particular, work should focus on surface modification of reinforcements; control of matrix-reinforcement adhesion; processing methods to control dispersement and alignment of reinforcements; hybridizing of nano- and microscale reinforcements; and integrating atomistic micromechanics and continuum modeling for predictive capability. Initial research into nanoparticles, such as exfoliated graphite and even cellulose microfibrils, indicate that such extremely low-cost systems offer multifunctionality enhancements comparable to carbon nanotubes; these opportunities should be fully explored, as should electrospinning of polymeric and other reinforcements that can produce submicron microfibrillar mats.

In contrast to the high probability of success in nanostructured PMCs, the potential for significant advances in nanocrystalline metals seems much more limited because the nanoscale grain size lacks stability with temperature changes and there are difficulties with processing and impurities. Much greater progress is likely with amorphous metals, where the best properties might be found in composite materials formed of an amorphous metal matrix with ductile dendritic phase reinforcements. Use of simple and fast thermoplastic forming techniques for these materials, the known glass transition temperature, and the thermal stability of the final amorphous composite merit further study. The panel recommends computational work to predict new potential alloy systems and create improved amorphous matrix composites (e.g., using SMA inclusions) to improve work hardening.

In health monitoring, use of wireless interrogation and networking technology should significantly improve remote access to localized structure information. Additional comments on health monitoring are found in the research priority section below, where NDI/NDE represent one of the major opportunity areas.

The self-healing capabilities of polymer-based composites have recently received wide attention due to work by researchers at the University of Illinois at Urbana-Champaign (White et al., 2001). Composite self-healing mimics nature's autonomic healing response. In polymers and polymer composites, the effect has been achieved by chemical triggers (catalyst) and microcapsules of fluid repair agent in the polymer matrix. When localized damage in the form of a microcrack reaches and bursts a microcapsule, the repair agent wicks into the crack, where it is polymerized by reaction with the catalyst. Tests on healed samples demonstrate no significant loss in material strength. This breakthrough in engineered self-healing for polymer-based materials deserves significant research emphasis, as it could lead to a wide array of self-healing polymers, composites, and adhesives for structures. It is critical to decrease to nearly instantaneous the time required for healing response.

Also under investigation for self-healing are metal-based composites, which include SMAs. At present, true self-healing has not been demonstrated; the focus has been on damage tolerance rather than self-healing. Nevertheless, intelligent material architecture has the potential to design materials with significantly improved damage tolerance and self-healing capacity by using SMAs that undergo phase transformation in the presence of the stress concentration from a crack. Also proposed is using SMAs for finite deformation mechanisms (as defined by the lattice transformation) for materials like amorphous metals used for work hardening.

There has been great progress in fire-safe polymers that can be used alone or as matrix materials for composites (Sorathia et al., 1997, 2001). Incorporating these materials into aircraft, spacecraft, ships, and other vehicles will decrease weight and improve safety. Fire-safe matrix materials will also lower the hazards of graphite fiber released into the air when aircraft built with such composites burn.

At least in concept, it is possible to design an almost unlimited number of composites with varying degrees of multifunctionality (tri-, tetra-, pentafunctionality, etc.) by employing both multifunctional matrices and multifunctional strengthening reinforcements. Some examples might include:

- A composite with a matrix containing microcapsules sensitive to thermal, electrical, or mechanical stress that on breaking would indicate the area of in-service damage to the material. Self-healing in the damaged area could then be accomplished by another family of microcapsules (as described above). The strengthening phases could be two different filaments that, in a conducting polymer matrix, could function as a battery.
- A microcellular structural foam used in the matrix. Some of these materials may be radar-absorbing, conducting, or light-emitting. With these, the strengthening filaments might have sonic sensing ability.
- Improved stealth coating. Stealth coatings today have several layers; roughness and wear are a critical problem. Composite technology, including nanoscale patterning, has the potential to create macroscale monolithic stealthy materials, removing the tremendous maintenance burden of layered structures.
- A photovoltaic military uniform cloth also containing kevlar body armor woven in at vulnerable places. During periods of sunlight the derived electricity could be used to maintain or replace charge in the batteries needed to power many of the devices to be carried by the infantryman of the future.

Adhesives and Coatings

Adhesives and coatings are used almost everywhere in military systems, most commonly to join components, seal, protect or insulate, produce patterns via photolithography, and in general to enhance the function and manufacturing of military systems. Adhesives are also used as matrix materials for all types of composites, including solid propellants and explosives as well as high-performance structural materials. Advances in the science and engineering of adhesion, including how such systems fail, will continue to yield improvements in adhesion performance and reliability. Substantial improvements over the next 20 years are expected; they will enable use of adhesives in ways not currently contemplated. Some likely advances and their possible applications are described here.

The ability to measure and characterize the chemical and topographical features of surfaces is important. Modern methods of chemical instrumentation make it possible to quantify the types of chemical groups active on surfaces. One can then design adhesive systems to bridge gaps between surfaces with different chemical functionality. These tools have also made it possible to modify the chemistry of surfaces to promote wetting or adhesion; previous methods, such as high-energy plasma treatments or oxidation, have not permitted much control and often damage treated regions. Today, chemical modification of surfaces is extensive; applications range from surface coatings to promote coupling to chemical reactions to alter the chemical makeup of surfaces. With modern lithographic methods, surfaces can be designed with topographic features that control wetting and promote adhesion. However, not all adhesion is chemical; some types of bonding are more physical, especially in porous materials and on surfaces with complex topography or woven fibers. In many of these systems mechanical interlocking is possible. Understanding of this type of adhesion is also advancing.

Progress in understanding of the failure of adhesive systems has come about largely from advances in the fracture mechanics of adhesives and improved methods of engineering analysis of adhesive systems. These methods improve understanding of the role of residual internal stresses, such as curing and thermal stresses, that can play a dominant role in many adhesive systems, as well as the response of the system to external loads and time. These advances in failure analysis, coupled with improved material characterization—including methods of representing aging—will continue to bring improvements in the reliability of all types of adhesion systems.

The stiffness of many adhesive systems can greatly affect the properties of composite materials. The maximum measured compressive strength obtainable for fiber-reinforced composite materials, caused by local elastic instability and often a limiting design feature, is dictated by the shear modulus of the matrix adhesive binding the system together, as it is in laminated sheet structures. Because most polymeric glasses, such as epoxies, all have about the same shear modulus and the spatial sizes are so small, attempts to improve matrix stiffness via composite reinforcement have not been pursued. With the advent of nanosized reinforcement, such as carbon nanotubes, it should be possible to substantially improve the stiffness of matrix materials by reinforcing them at the nanoscale level, thereby increasing the measured compressive strength of the fiber-reinforced composite.

The concept of repairable adhesives is hovering on the horizon: Adhesives are commonly used in chip attachment in microelectronics. Modern chips are so expensive that their repair in complex circuits is not only justified but necessary. This has led to the need for adhesive systems that can be repaired by selectively destroying the bond via chemical or thermal means so that damaged parts can be removed and replaced. Embedded sensors that can help quantify the life of adhesive systems, coupled with design of repairable systems, should greatly add to the reliability and use of adhesives.

Significant advances in high-temperature adhesives and sealants are expected with the evolution of hybrid organic/inorganic materials that will also extend the utility of adhesive systems.

RESEARCH AND DEVELOPMENT PRIORITIES

This chapter discusses the most notable DoD structural materials needs. From these, the panel has extracted four broad R&D priorities. Investment in these priority areas should provide enabling materials technologies to DoD for use by 2020.

Materials Design Assisted by Computation

In the design of materials assisted by computation, the goal is to implement new materials by integrating constitutive models into a framework that employs FEM calculations. The first-level potentials (electron, atom, dislocation, and microstructure) have to be described to yield time, temperature, and size-dependent models of material behavior for all classes of materials, monolithic and composite. This effort will require more precise understanding of physical phenomena and better computer equipment that can extract the important data from a calculation and also analyze these first-level data.

A successful effort would predict possible material properties before development costs are incurred and predict material properties and behavior so that the cost of characterization (e.g., temperature dependence of polymer-based materials) could be reduced. This would not only reduce costs but also accelerate material development, especially the introduction of new materials and materials systems into DoD systems. The best approach may be to pick a sample application and then assemble teams of specialists—computational materials experts, structural analysis experts, systems experts, and materials scientists—to demonstrate atoms-to-structures design.

Service-Induced Material Changes

As a result of the improved understanding emerging from computation-assisted design of materials it will become possible to use these materials to design better structures. Thus, the second opportunity identified by this panel involves service-induced material changes or how to use materials. The rationale for this concept is that materials and structures ought to be designed based on material functionality in actual service rather than on the material's initial properties. This means understanding history-dependent properties and performance evolution. The science base for the use of materials must therefore be expanded, and nonequilibrium structures and materials must be better understood. This will be the basis for a science-based constitutive theory that is explicit about the extensive variables of the material.

Among the benefits of such an approach would be history-based reliability prediction for individual components and replacement of "factors of ignorance" in design with deterministic model-based sensing and prediction. Other benefits are robust multidisciplinary virtual design environments that reduce cost and speed design, testing, prototyping, manufacturing, and deployment. The results would be reduced vehicle weight, enhanced performance, and fuel savings.

To be successful, this effort must be complemented by efforts to predict in-service loads, which in turn would require new model-based life-cycle sensing and design methodology based on computational and phenomenological models of material state evolution.

Multifunctional Composite Materials

R&D on composite materials needs to advance in several areas simultaneously. The most important general advances needed are in the processing and scale-up of material combinations. Advances needed for MMCs and CMCs include adequate exterior coatings for nonoxide composites, fiber coatings to prevent fiber-matrix chemical diffusion at higher temperature, and interface control to tailor bonding between filaments and matrix. Advances needed for PMCs incorporating high-performance polymer fibers include improvements in compressive strength and matrix stiffness. Other necessary advances for composites are in the area of structural multifunctionality, for purposes of NDI/NDE integration as well as adding thermal management, actuation ability, etc. This exciting possibility employs some of the multifunctional monolithic polymers identified in Chapter 6 as the matrix phase in a structural composite.

Benefits of this effort by 2020 would be improvements of 20-25 percent in strength, toughness, stiffness, density, environmental resistance, and high-temperature capability, leading to enhanced mobility, maneuverability, survivability, and transportability of DoD systems. New multifunctional composites would increase warfighter capabilities while integrating formerly discrete systems into a single package.

Integrating Nondestructive Inspection and Evaluation into Design

As a complement to composite material development, this panel believes that there are opportunities for integrating NDI/NDE into the original design of both materials and structures. This would allow for continuous health monitoring of all new structures. Sensors integrated into structure would have to be very small, so many new types of sensors will be needed, as will be portable advanced sources, such as X-ray and neutron sources, to allow field evaluation of structures. Some sources should be incorporated into internal structure wherever the structures will be difficult to examine with an outside source.

Salient results of this effort would be real-time recovery of information from material (especially multifunctional) systems that can be used to evaluate safety, reliability, need for maintenance or replacement, or remaining strength and life. This information would support management decision-making at all stages of engineering design and development. It would depend on new methodologies for recovering, interpreting, and using information from integrated sensors or sensor arrays in material systems, and the actual recovery of such information.

REFERENCES

American Society for Metals. 1961. Metals Handbook, 8th Edition. Vol. 1. pp. 44-48. Metals Park, OH: American Society for Metals.

Antolovich, S.D. 1992. Superalloys 1992, Proceedings of the Seventh International Symposium on Superalloys, September, Champion, PA. Warrendale, PA: The Minerals, Metals & Materials Society.

Defense Science Board, Task Force on Improving Fuel Efficiency of Weapons Platforms. 2001. More Capable Warfighting Through Reduced Fuel Burden. Washington, DC: Office of the Under Secretary of Defense for Acquisition, Technology and Logistics, January. Available online at <www.acq.osd.mil/dsb/fuel.pdf>. Accessed May 21, 2001.

Department of the Army, TACOM. 2001. Proceedings of the 12th Annual U.S. Army Ground Vehicle Survivability Symposium, March 26-29, Monterey, CA.

Kissinger, R.D. 1996. Superalloys 1996, Proceedings of the Eighth International Symposium on Superalloys, September, Champion, PA. Warrendale, PA: The Minerals, Metals & Materials Society.

National Research Council. 1998. Ceramic Fibers and Coatings: Advanced Materials for the Twenty-First Century. Washington, DC: National Academy Press. Available online at <www.nap.edu/catalog/6042.html>. Accessed November 14, 2001.

Pollock, T.M. 2000. Superalloys 2000, Proceedings of the Ninth International Symposium on Superalloys, September, Seven Springs, PA. Warrendale, PA: The Minerals, Metals & Materials Society.

Schaffer, J.P., A. Saxena, S.D. Antolovich, T.H. Sanders, Jr., and S.B. Warner. 1955. Pp. 614-619 in The Science and Design of Engineering Materials. Chicago: Richard D. Irwin.

Seely, F.B., and J.O. Smith. 1955. Advanced Mechanics of Materials. New York: John Wiley & Sons.

Sims, C.T., N.S. Stoloff, and W.C. Hagel. 1987. Pp. 27-32 in Superalloys II. New York: John Wiley & Sons.

Sorathia U., R.E. Lyon, R.G. Gann, and L. Gritzo. 1997. Materials and fire threat. Fire Technology 33(3):260-275.

Sorathia U., T. Ohlmiller, R. Lyon, J. Riffle, and N. Schultz. 2001. Effects of fire. Pp. 100-120 in Gap Analysis for Durability of Fiber Reinforced Polymer Composites in Civil Infrastructure. Washington, DC: Civil Engineering Research Foundation.

White, S.R., N.R. Sottos, P.H. Geubelle, J.S. Moore, M.R. Kessler, S.R. Sriram, E.N. Brown, and S. Viswanathan. 2001. Autonomic healing of polymer composites. Nature 409(6822):794-797.

CHAPTER FOUR

Energy and Power Materials

CHAPTER SUMMARY

The flow of energy and power forms the lifeblood of military systems. The Panel on Energy and Power Materials was formed to examine advanced materials and processes in this area. Every other panel of the full committee also contributed to identifying challenges to be met in energy and power materials.

DoD needs for energy and power materials are many, among them:

- Batteries for energy storage, from small portables to large shipboard units,
- Capacitors for storage and release of pulsed power,
- Fuel cells for efficient direct conversion of chemical to electrical energy for platform power,
- Photovoltaics for harvesting energy from the environment,
- Explosives for enhanced and tailorable lethality, and
- Microturbines for powering small unmanned aerial vehicles (UAVs).

Because this area is so broad, a comprehensive study was not possible. Instead, the panel identified key materials aspects of each major application and derived broad themes for materials research. Areas identified were those where DoD funding would be needed due to the lack of commercial interest, highlighting differences in commercial and military requirements. Major research themes identified are

- Nanomaterials science and engineering for control of structure and properties at the nanoscale;

- Engineering interfaces and surfaces in materials by tailoring material structure to optimize rate and extent of reaction processes;
- Advanced energy storage and conversion materials;
- Tools for accelerated, systematic materials discovery, both analytical and experimental, e.g., computational materials science and combinatorial materials science; and
- Materials as the foundation for systems, based on systems approach to materials in which the entire suite of material properties relevant to a class of applications is identified as early as possible to avoid point solutions.

Successful pursuit of these themes will provide numerous benefits to the DoD, including:

- Reduced development time and cost;
- Increased energy density in storage devices, with associated weight reduction;
- Improvements in lethality of munitions, including increases in range and payload;
- Practical energy-harvesting devices to allow capture and storage of solar radiation in the battle area; and
- Reduced weight of energy and power systems, which will reduce soldier and system payload.

Additional information on DoD needs, research themes, and benefits is provided in the body of this chapter and in Appendix D.

INTRODUCTION

Energy and power are ubiquitous in DoD platforms. To satisfy the broad-ranging charter given to this committee, a separate Panel on Energy and Power Materials was appointed to treat such subareas as batteries, fuel cells, and energetic materials. Other panels addressed specific energy and power areas that fell within their purviews. As an example, biological materials approaches to power and energy (e.g., ATP analogs) were covered by the Bioinspired and Bioderived Materials Panel (see Chapter 7). Figure 4-1 identifies which panels covered which energy and power materials subareas.

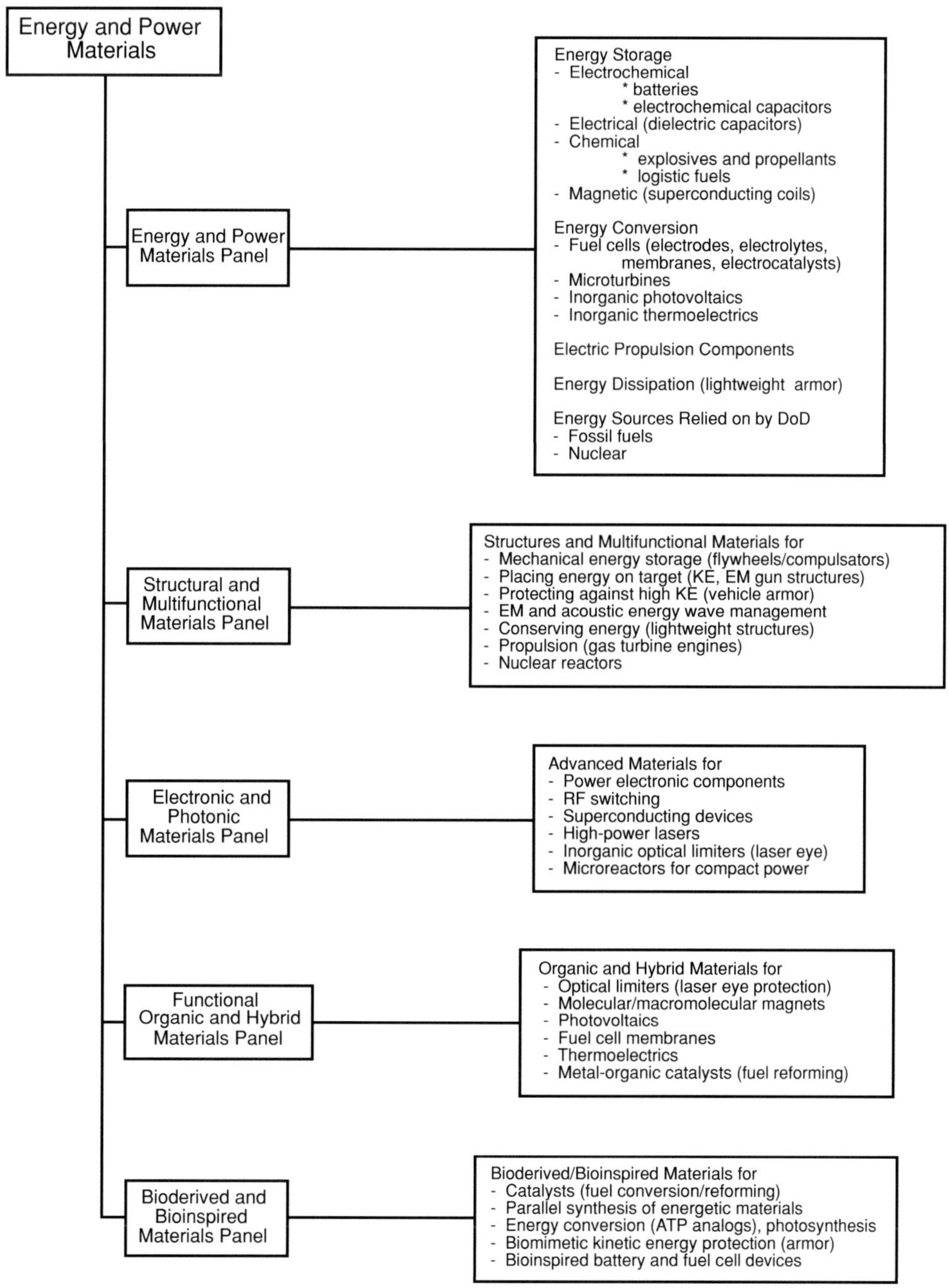

FIGURE 4-1

Energy and power materials addressed by panels of the committee.

Under energy storage, the Panel on Energy and Power Materials examined materials challenges for electrical, electrochemical, chemical, and magnetic storage of energy. This area included materials for improved batteries and for capacitors, both electrical and electrochemical, as well as chemical energy storage in the form of explosives, propellants, and fuels that could provide a significant advantage to the military in 2020. Explosives and propellants are developed almost solely by government entities or their industrial contractors. While the scientific literature was examined, the information available likely reflects work already known to DoD. In this case, the most important contribution of the panel may have been to identify opportunities that are not being pursued aggressively due to limited budgets and a current focus on immediate needs and near-term payoff. Also, while mechanical energy storage (particularly in flywheels/compulsators) is important to DoD, the associated issues are largely structural (see the work of the Panel on Structural and Multifunctional Materials reported in Chapter 3).

The panel also examined challenges for efficient conversion of energy from one form to another. The most important component in this category, for both small-scale and large-scale energy conversion systems, is the fuel cell. The panel also examined materials for inorganic photovoltaic (PV) devices—organic PV materials were covered by the Functional Organic Materials Panel (see Chapter 6). The Energy and Power Panel also considered the conversion of chemical energy into thrust in very small-scale applications via microturbines, while the Structural Materials Panel addressed materials for larger gas turbine engines.

Materials challenges to fielding advanced weapons, i.e., placing energy on-target, were considered by the panel only in the context of energy storage or conversion, because many of the components of these weapons use materials within the purview of other panels. Materials for advanced high-power lasers, for instance, were addressed within the Panel on Electronic and Photonic Materials (see Chapter 5). Similarly, the problems with materials for advanced gun tubes and for electromagnetic launch rails primarily relate to erosion, wear, and structural integrity, again more aligned with the mission of the Panel on Structural Materials. However, the Panel on Energy and Power Materials did consider the development of materials for energy storage at high levels that might accelerate associated applications.

The need to dissipate concentrated energy and protect systems against its effects is ubiquitous and was addressed by several panels. Materials challenges for effective kinetic energy dissipation (e.g., novel armor) were addressed by the Panel on Energy and Power Materials (transparent armor and body armor) and the Panel on Structural Materials (integrated structural armor and protection of large platforms). In both cases, information on potential performance improvements from combining new materials with new design approaches was difficult to obtain, because that information was often classified. Materials challenges for hardening against other forms of energy (e.g., acoustic, thermal, radar) are discussed in Chapter 3 in the examination of multifunctional materials.

DoD dependence on natural energy sources was considered, in relation to DoD platforms rather than fixed installations. Accordingly, less emphasis was placed on sources such as water and wind power. Solar energy was recognized as having some potential for use in energy conversion devices for platforms and individuals. The panel did look at materials that could significantly improve DoD's ability to harvest energy from alternative sources, thereby improving field power-generating capability while decreasing the logistics burden of supporting expeditionary forces. Potential shifts in the reliance of DoD on natural sources such as fossil fuels were examined only briefly because the need for changes in U.S. energy sources, particularly a shift from fossil fuels to more rapidly renewable resources, has been well documented. If petroleum-based fuel prices were to accelerate dramatically, a substantial U.S. initiative to find and apply alternative energy sources would likely result, with DoD being one major beneficiary. The potential need by 2020 for fossil fuels, as well as future dependence on nuclear power, are also discussed in this chapter.

DOD NEEDS FOR ENERGY AND POWER MATERIALS

Virtually every DoD system requires energy and power to function. Attaining higher energy and power levels in a smaller package is a continual goal in development of new materials, manufacturing processes, and design approaches. Examples are

- Battery materials (small, ultralightweight power cells to large submarine batteries);
- Fuel cells (direct conversion of chemical to electrical energy across a range of applications);
- Capacitors (pulsed power for electromagnetic launch of projectiles, aircraft, etc.);

- Explosives and propellants (insensitive yet possessing higher energy density);
- Advanced individual body armor (to offset the increased lethality of individual weapons);
- Novel power system components (for all-electric weapons platforms); and
- Components for harvesting energy from the environment (e.g., photovoltaics, thermoelectrics).

These examples are illustrative only; an exhaustive list of energy and power applications would require enumeration of every DoD system.

SPECIFIC AREAS OF OPPORTUNITY

In the sections below, DoD needs for energy and power materials are examined in specific categories:

- Energy storage,
- Energy conversion,
- Electric power generation and transmission,
- Kinetic energy dissipation, and
- DoD reliance on energy sources.

Additional information on materials for specific applications is provided in Appendix D.

Energy Storage

In any system, the rates of energy storage and consumption, combined with the total energy available, determine how long the system will be effective, i.e., the system run time. From this standpoint, the ideal system would have an infinite reservoir of energy (or infinite rate of energy storage) and a negligible rate of energy consumption. While real systems never approach these limits, there is nevertheless a continuing drive for approaches that can store large amounts of energy in small packages and use it efficiently. In materials terms, advanced energy storage requires materials having high volumetric and gravimetric energy densities.

Although storing electrochemical energy in batteries may be most familiar, energy can be stored in many forms: chemical (in fuels and explosives); mechanical (flywheels); electrical (capacitors); thermal (solids, liquids, or gases); magnetic (superconducting circuits); and potential (reservoirs). This report focuses on energy storage in the forms most traditionally associated with DoD platforms and systems.

The future of materials science as it relates to energy storage systems is exciting. For electrochemical systems, advances in nanomaterials can yield ionically conducting electrolytes that can provide both high power and high safety. New discoveries promise high-energy electrode materials and more stable electrolytes that will greatly increase the energy content of current batteries for applications ranging from the dismounted soldier to submarines. Tailored interfaces will increase the lifetime of electrochemical power systems and make them capable of power pulses well beyond what can be delivered today, including in radio burst communication from ground to satellite. Similarly, electrical energy can be stored in dielectric capacitors for use in pulsed power systems for high-energy lasers, railguns, and other advanced weapons.

Materials that are extremely mass-efficient in storing chemical energy are important for DoD. These "energetic materials" will enable the lighter, more lethal force structure envisioned by the Army beginning now and continuing through 2020.[1] A driving goal for much of the DoD-wide system transformation is a shift to smaller platforms with increased lethality.[2] Higher-energy-density explosives and propellants combined with precision targeting systems translate into smaller warheads, enhanced penetration, longer range, and reduced ammunition logistics support.[3] In addition, novel propellant materials may be able to reduce launch signature, thus increasing survivability by reducing detectability.[4] This class of materials includes chemical fuels for platform mobility, which now account for some 70 percent of war tonnage shipped to combat locations

[1]Andrews, M., "Army Vision and S&T: Accelerating the Pace of Transformation," paper presented to the Committee on Materials Research for Defense After Next, National Research Council, Washington, DC, February 15, 2000.

[2]Ullrich, G.W., "Advanced Energetic Materials: Introduction and Overview," paper presented to the Committee on Advanced Energetic Materials and Manufacturing Technologies, National Research Council, Washington, DC, July 31, 2001.

[3]Lannon, J., "U.S. Army Energetic Materials," paper presented to the Committee on Advanced Energetic Materials and Manufacturing Technologies, National Research Council, Washington, DC, July 31, 2001.

[4]Finger, M., "Energetics Survey," paper presented to the Committee on Advanced Energetic Materials and Manufacturing Technologies, National Research Council, Washington, DC, July 31, 2001.

(Potomac Institute, 2000) and which are absolutely essential for a wide range of platforms. In addition, energy stored in magnetic fields can reduce the effects of short-term power disruptions. Novel superconducting materials that may increase storage efficiency and decrease the size and complexity of the storage system are discussed later in this chapter.

What energy storage technology will best meet DoD needs? A single answer is not possible, because all power sources have their "sweet spots" (areas of best application), and this panel envisions that a range of energy storage (and conversion) technologies will be important. For example, low-power needs (usually for electronic devices) are optimally met by storage in batteries. Other applications require other technology choices, from storing energy in a highly concentrated, liquid fuel form for use in internal combustion engines to storage in capacitors for high-power directed energy weapons.

The National Academies have published two studies of the use of power sources for defense applications, *Energy Efficient Technologies for the Dismounted Soldier* (NRC, 1997) and *Reducing the Logistics Burden for the Army After Next* (NRC, 1999). Both highlight the importance of developing advanced technology for power sources, including batteries, for DoD applications. This study builds on these documents.

The sections that follow discuss novel materials and processes for use in storing energy. Additional information is provided in Appendix D.

Electrochemical Energy Storage

Background: Batteries and Electrochemical Capacitors Batteries and electrochemical capacitors are energy storage devices that convert chemical into electrical energy and are particularly suited to provide the energy to power electrical devices. They cannot be easily replaced, and are very likely to be an essential element in any future military power applications. Their simplicity (no moving parts), reliability, and wide power capability make them attractive as stand-alone power systems or as enabling elements in hybrid power system configurations. Additionally, batteries are the best option for stealthy operation. They produce power with no noise, no heat signature, no intake of oxygen, and no exhaust gas. Special Operations Forces will always need better batteries.

Moreover, while there are more efficient ways to store energy, few other devices can match the attributes of batteries and electrochemical capacitors. Of particular importance is the fact that very few other power sources have a wide enough dynamic range to efficiently deliver power to follow a dynamic load over three to four orders of magnitude. Several types of batteries have this feature, which is of particular importance in very high power applications and in hybrid power sources (ones that have a tailored power source with a *high energy* component and a *high power* component designed to work together for optimum performance).

Batteries are electrochemical cells that have two electrodes, an anode and a cathode. During discharge, the anode is oxidized and the cathode is reduced. The electrolyte (solid or liquid) is ionically conductive, allowing ions to be transported between the electrodes. In these systems, the electrons are driven through an external circuit to power electronic devices. Batteries can be disposable (a primary cell) or reusable (a rechargeable or secondary cell). Electrochemical capacitors also have two electrodes separated by a separator having an ionically conductive electrolyte, but their energy can be composed of double layer capacitance (DLC) or it can be stored in Faradaic processes (much like a battery). Either way, they are rechargeable devices designed to have a very high cycle life. Like conventional capacitors, electrochemical capacitors have higher power and lower energy than rechargeable batteries. Figure 4-2 demonstrates the comparison.

DoD system energy and energy storage requirements vary enormously, ranging from tiny batteries to power man-portable communication systems to huge systems to power submarines. Diversity in capacity, physical size, weight and shape, drain rate capability, thermal performance, cycle life, shelf life, and cost make it impossible to define a single optimal battery. The need for a systems approach to finding new materials to match power source to application for DoD will be a continuing theme of this chapter.

Both batteries and electrochemical capacitors can deliver energy over a wide power range, making them well suited to hybrid power systems. A timely example is the hybrid electric vehicle (HEV), which contains both an internal combustion engine (ICE) and a battery pack. The ICE can operate at peak efficiency to charge the battery pack, which in turn provides power for propulsion. Properly designed, one can have the benefits of both systems—the high energy content of a liquid fuel and the high power of a battery.

Military-Unique Requirements There are specific military requirements that limit the use of commercial off-the-shelf (COTS) power sources, among them differences in required power and energy levels, in temperature operating range, in the need for shock resistance, in the requirement

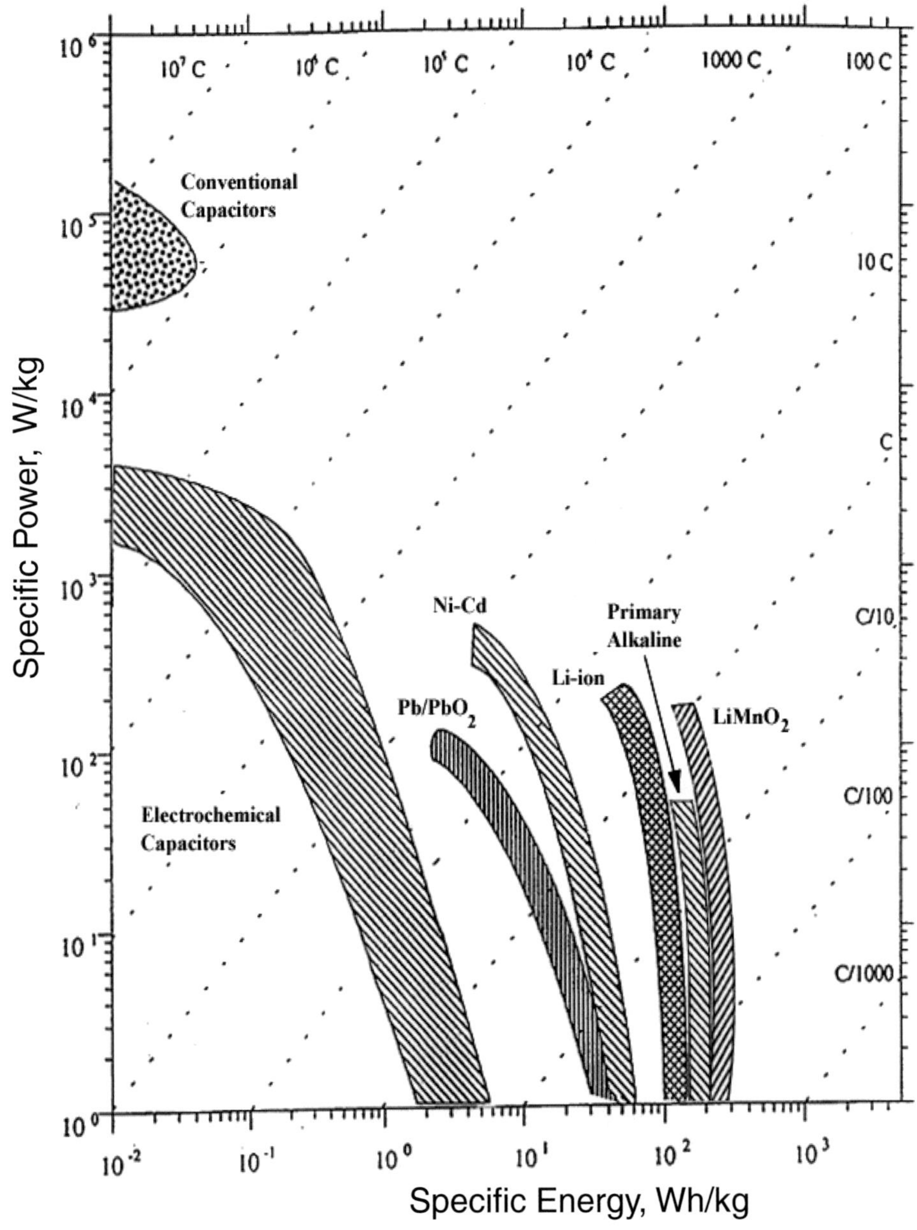

FIGURE 4-2

Ragone plot comparing nominal performance of batteries, electrochemical capacitors, and dielectric capacitors. SOURCE: Reprinted from Christen and Carlen (2000), Copyright 2000, with permission from Elsevier Science.

for longevity, and in reliability. Except in HEVs, high-power batteries are generally not needed in the commercial marketplace, and even the HEV is not in a performance realm that can meet future DoD requirements. Operation in a broader temperature range and under more rigorous environmental extremes will be necessary. The military has a greater need for increased shelf life and lower self-discharge rates than typical commercial batteries offer. Furthermore, commercial devices are not held to the same reliability standards as those required by the military.

The reliability driver for commercial power sources is to minimize warranty expenses; the reliability driver for future DoD systems is mission success and soldiers' lives. Such differences are typified in Figure D-2 on naval batteries (see Appendix D). Materials R&D will be necessary to help fill these gaps.

Research Challenges and Materials Opportunities To translate technology needs into materials needs, one must map the system requirements to the materials domain. From a materials perspective, battery performance is dictated by materials selection and stability (Salkind, 1998; Doughty, 1996). Intrinsic properties of the active electrode materials determine the cell potential, capacity, and energy density. The stability of interfaces between reactive materials dictates calendar life and cycle life (in rechargeable systems); safety depends on the stability of materials. Because battery performance depends heavily on tailored stable interfaces, it is here that solutions will be found.

Nanomaterials Nanostructured materials are of wide interest in research programs today. They hold the promise of tailored and engineered materials that were inconceivable a few years ago. Continuing these advances will provide important benefits for power sources if they can be reliably produced and stabilized so that their nanostructure (and associated properties) remain relatively constant over extended use. Specific areas that will benefit are tailored electrode materials and electrolytes.

New processing methods for advanced micro- and nanostructures is one fertile area. While high-power electrodes require high-surface-area materials, processing often does not allow the technology to take full advantage of the material. Nanostructured materials combined with a one-step deposition process or a self-assembly approach to fabricating the electrode may overcome these limitations. Tantalizing initial steps are already being taken; more effort should be devoted in the next two decades to make such nanostructural processing a reality.

Similarly, thin-film processing techniques have been shown to overcome certain failure mechanisms (using glassy electrolytes to avoid lithium dendrite formation—see below) and have extended the cycle life by over two orders of magnitude (Bates, 2000). Pulsed laser deposition of active materials holds promise (Singh et al., 2000), as do thin-film/thick-film deposition techniques, such as plasma spray and plasma-assisted chemical vapor deposition (CVD). Further extensions of these methods, combined with wholly new approaches, are required. An example is fabrication of the battery layer by layer with solid-state materials. In this approach, the cathode and anode could be surrounded by materials that have, respectively, cathodic and anodic stability. The increase in calendar and cycle life and in high-power delivery could be dramatic.

Tailoring morphology and texture of the electrode material to form high-surface-area composites is another promising area. Mesoporous materials in which nanometer-sized domains are interconnected in a porous electrode can provide up to 100 percent greater capacity than polycrystalline nonporous V_2O_5 powers, with exceptionally high rate capability (Dong et al., 2000). This is believed to result from increased electrode capacitance, not just larger Faradaic capacity. High-capacity anodes are made from nanostructured alloys or metal/oxide composites. In both cases, large-volume expansion upon charge can be accommodated with in-situ formation of nanostructured electrodes on the first charge, resulting in higher capacity and capacity retention on cycling (Poizot et al., 2000). Nanostructured electrochemical capacitor electrodes providing much higher capacitance (nearly 700 F/g) are also being investigated (Pang and Anderson, 2000). Metal oxides like RuO_x have very high capacitance but are quite expensive. Producing other metals with high redox activity in a nanostructured state could heighten capacity and improve volumetric efficiency for supercapacitor electrodes.

Electrolytes for polymer systems have been a huge challenge for technology developers. The "polymer batteries" that are now becoming commercially available employ gel electrolytes that have more in common with liquid electrolyte systems than with true polymer systems. Shape flexibility, lower weight, negligible magnetic signature, and small form factor are important advantages. Nanostructured electrolytes could provide breakthrough technology. For example, organic/inorganic composites hold the promise of creating new ionically conducting channels in polymer systems that are not limited by the current mechanism responsible for ionic mobility—segmental chain motion in polymers. Self-assembled nanostructures have already been described for use in electrolytes (Bronstein et al., 2001). Improvement in conductivity and transference number have been observed with nanostructured clays in polyethyleneoxide (PEO) (Riley et al., 2000) as well as nanoparticulate oxides in PEO (Croce et al., 1998).

Composites can alter the mechanism of ionic transport, suggesting the possibility of high conductivity, with ionic transport near unity of the "working ion" (e.g., Li^+) (Sata et al., 2000). Advances in this area will have application in high-power batteries because they will have low internal impedance and be able to operate over a wide temperature range. For example, if ionically conducting perovskites (like $(Li_{1-x}La_x)TiO_3$) could be self-assembled in a polymer film to form ionically conducting channels in a thin-film electrolyte, they might provide room-temperature ionic conductivity as well as serve as a robust separator to enhance safety.

Interface Science Interfaces lead to many problems in batteries and electrochemical capacitors. For example, calendar life deteriorates in lithium ion rechargeable batteries when resistive decomposition products are deposited at the cathode. The power of electrochemical capacitors would improve if the interfaces (where charge is actually stored) were engineered to provide electrochemical stability as well as optimum charge storage. Of major interest to DoD applications would be improved performance at higher frequencies. Little work is now funded on interfacial chemistry as it relates to power sources.

An example of a practical need is stabilizing lithium metal electrodes in rechargeable batteries. The payoff is 2-3 times greater energy density, but the problem is safety. Repeated charging and discharging produces Li dendrites that rob capacity and create a safety hazard by producing a fine pyrophoric Li powder. Tailoring the interface with new additives or by creating ionically conducting nanostructured polymeric or glassy layers on the metal surface could profoundly affect the safety of high-energy-density batteries. Examples are thin oxide layers that are ionically conducting. Currently, lithium phosphorous oxynitride (LiPON) glass is used as the electrolyte for planar batteries. This material can be improved. Other materials might be developed that are ionically conducting like

$(Li_{1-x}La_x)TiO_3$ perovskites, that are stable in these environments, and that stabilize the Li surface by forming a graded interface.

New Electrode and Electrolyte Materials Every few years, the materials community is surprised by the discovery of new materials with unexpected properties that prove extremely useful. Examples are high-T_c superconductors and new electrode materials. Research into cathode materials has centered on layered transition metal oxides (e.g., $LiCoO_2$ and related compounds). A new material, $LiFePO_4$, has similar capacity (at low rates) and dramatically improved safety at a fraction of the cost of existing materials, but presently it has very poor electronic conductivity, making it a laboratory curiosity. Investments in new electrode materials will likely yield other new materials with improved performance, new electrolytes with improved conductivity, and new processing techniques (enabling new concepts in microbatteries).

Metal/air and metal/water batteries have some of the highest energy content of any battery systems but materials problems have prevented their widespread use in large applications (Hamlen, 1995). Water management and carbonation of alkali electrolytes remain roadblocks. New polymeric membrane materials that address these problems could have a major impact, as would semipermeable membranes that transport moisture and exclude CO_2. Advances in tailored membrane materials and controlled porosity and surface chemistry could be targeted to this application.

Computational and Combinatorial Materials Science "Materials by design" is an appealing approach because it can dramatically shorten the time to develop advanced materials. While it can apply to nanostructured materials, it is a more general concept that may be used to tailor materials microstructure. Therefore, computational materials science should be an integral part of any systematic materials discovery effort. In particular, advanced computational techniques that model performance of materials in active electrochemical environments can point the way to new and potentially revolutionary battery materials.

Similarly, combinatorial chemistry should be employed where possible. It has been used with success in the biological sciences and in drug discovery, and it is appearing in the materials science arena. Initial work is now underway on preparation of alloy anode materials using this approach, but it is limited by the lack of rapid screening techniques. Investments in this area will dramatically shorten the time to realization of electrochemical materials that can have a beneficial effect on DoD power sources.

Materials for Dielectric and Magnetic Energy Storage

Appendix D summarizes material requirements and prospects for storing electrical energy as separated charge in a dielectric material like a ceramic or a polymer. Compared to other forms of electrical energy storage, such capacitors are lower in energy density but higher in power density, and they are more readily cycled at high frequencies. Pulsed power applications for future DoD systems may require from 0.4 to 40 MJ/discharge, with an average instantaneous power of 0.5 to 7 Gwatt.[5]

Future ceramic-based power capacitors will require novel formulations and processing methods for tough, defect-free, low-loss ceramic and glasses for use in configurations that are scalable to different sizes to meet different application needs. Recent advanced materials approaches have focused on improved mixed-oxide perovskites, diamond-like carbon, and high dielectric constant ceramics produced using thin-film techniques. Here as elsewhere, computational chemistry, combined with the ability to control structure down to the nanoscale, should pay tremendous dividends. Recently, polymer blends having high dielectric strength and dielectric constant (>8.0) combined with low loss have been produced through physical alloying of polymers. Similarly, PVDF copolymers with nanodomains generated by irradiation have been produced, which suggests the possibility of attaining very high specific energy storage levels (20 kJ/kg) compared to current materials (Clelland et al., 2000).

The use of electromagnetic fields to structure composites that contain micron and submicron particles should also be examined. This approach has already resulted in materials with unusual properties. Simulations of materials made using this principle have shown that field-structured dispersions of high-dielectric-constant nanoparticles of uniform size possessing organic capping layers may greatly increasing the dielectric constant of composite polymeric thin film.[6] Whether this approach improves upon or detracts from other properties essential to power capacitors remains to be determined. Also, though the material properties of numerous potential high-dielectric-constant materials have been evaluated in the laboratory, few have been evaluated as part of a system or subsystem,

[5]McNab, I.R., University of Texas at Austin, private communication, August 26, 2002.
[6]Phillips, J., Sandia National Laboratories, private communication, March 30, 2001.

where questions about how material properties translate into device performance could be answered relatively inexpensively and straightforwardly.[7] This subsystems-level examination should be considered as it would provide DoD the ability to identify potential capacitor material candidates rapidly and with greater precision than is now possible.

Magnetic storage is another way to ensure a ready reservoir of energy. Commercially available superconducting magnetic energy storage (SMES) systems can instantaneously release stored electrical energy to mitigate short power outages or voltage sags, whether AC or DC. The current systems are large and located in fixed installations or sized to fit into the trailer of an 18-wheeler. Materials advances in high-T_c superconducting materials will enable these to operate at temperatures above 77 K, permitting a switch from liquid helium to the more readily available liquid nitrogen. Miniaturization of high-power conversion systems using advanced materials combined with smaller cooling systems will lead to SMESs that are more easily deployed, providing stable, highly reliable, and readily available power for sensing, control, communication, and data processing on the battlefield or on board all-electric ships. To take advantage of these capabilities, materials R&D effort should concern itself first with increasing the T_c and J_c of HTS ceramics and ceramic composites; increasing processability of these new materials while decreasing their cost; exploring higher T_c and J_c of MgB_2 and its alloys; and using combinatorial experiments to identify new HTS materials.

Chemical Energy Storage

"Energetic materials" is a collective term for military high explosives, propellants, and pyrotechnics; it is often used as a synonym for "military explosives" (DoD, 1999). In this chapter, this definition is expanded to include chemical fuels, such as diesel and jet fuel—an important class of materials for powering military platforms. Other fuels, including options for use in fuel cells, are discussed in the section on materials for energy conversion.

Whereas chemical fuel development has been led largely by commercial firms, development of explosives and propellants has been funded almost exclusively by the military, and information about it is often closely controlled. The panel used information obtained from unclassified govern-

[7]Sarjeant, J., State University of New York, Buffalo, private communication, January 29, 2002.

mental sources, publications appearing in the scientific literature in specific subareas, such as nanotechnology. The panel notes that research in explosives and propellants has concentrated on insensitive munitions over the years and that funding for new materials has not kept pace with the potential for improvement. DoD has recognized this with the recent formation of the Advanced Energetics Initiative, designed in part to rectify this imbalance. Appendix D provides some background information on development of explosives and propellants and their potential for improvement. The remainder of this section discusses the associated research opportunities.

Propellant and Explosive Materials DoD needs energetic materials that are insensitive (i.e., resistant to accidental explosion) and that offer higher energy densities and the ability to tailor energy release for different uses. Tools exist that can speed the development of energetic materials while reducing the cost and time-intensive testing necessary to verify their performance. These include (1) computational materials science in materials design to help identify promising new molecules and formulations; (2) novel syntheses, including the use of combinatorial chemistry for materials discovery; (3) improved processing methods; (4) new characterization techniques; and (5) advanced modeling techniques to predict the effects of testing. These are key enabling tools referenced repeatedly in the panel's deliberations.

Because this study was not classified, it may not take into account areas known to DoD that have high promise. Key materials-related research areas identified by the panel as having high promise are

- High nitrogen/polyatomic nitrogen energetic molecules,
- Nanotechnology for energetics,
- Reactive intermolecular materials, and
- Nuclear isomers.

Polyatomic nitrogen compounds have the potential to yield propellant materials with double or triple the specific impulse I_{sp} of hydrazine, and also have potential for enhanced explosives.[8] These compounds include

[8]Danen, W.C., "Los Alamos Advanced Energetics Programs," paper presented to the Committee on Advanced Energetic Materials and Manufacturing Technologies, National Research Council, Washington, DC, August 1, 2001.

unusual nitrogen complexes, such as $N_5^+AsF_6^-$ salt, that yield energy by producing nitrogen molecules. The synthesis of this compound, which is stable but produces high heat of formation, was announced in 1999 but it needs additional work to bring it to reality.[9] The real payoff, however, may be the discovery of additional molecules of this family. For example, salts employing polyatomic species that include N_4 or N_8 ions might allow for a 5-6 times reduction in mass for the same payload mass in missiles, making for huge increases in performance. However, proof of concept would require considerable cost and time to demonstrate, to be followed by development of practical processes for processing these materials in quantity.

Nanoscale aluminum powders produced by such processes as gas condensation are now becoming available in commercial quantities; they make it possible to tailor propellant burn rate, allowing for increases in projectile range or in super-thermite reaction velocity.[10] Burn rate increases of 30-fold over standard composite rocket propellants have been identified, and more advanced nanocomposites are possible. Although the outcome is still limited by thermodynamics, intimate mixing at the nanoscale may lead to a 40 percent increase in energy density over currently used materials. This should provide higher payload and perhaps decrease munition sensitivity.

Of particular interest in nanocomposite energetic materials are factors like nanoparticle size, size distribution, morphology, surface chemistry, and composition. These variables have all been shown to be important in determining mechanical, barrier, and other properties of nanocomposites (Kornmann et al., 1998; Messersmith and Giannelis, 1995) and will likely control the contributions of nanoscale components in energetic materials as well. However, any effort to examine nanocomposites for energetic materials will likely require sorting through the materials currently used in propellants at larger sizes as well as examining a myriad of alternatives for combination at the nanoscale. Such an effort could benefit greatly from computational materials science and combinatorial chemistry techniques.

While many explosive molecules are intramolecular materials, metastable intermolecular composites (MICs) are clusters of reactant materials

[9]Christe, K.O., and W.W. Wilson. Paper presented at 14th Winter Fluorine Conference, ACS Division of Fluorine Chemistry, St. Petersburg, FL, 1999.

[10]Dye, R., "Breakthroughs in Nano-Engineering," paper presented to the Panel on Energy and Power of the Committee on Materials Research for Defense After Next, National Research Council, Irvine, CA, October 10, 2001.

that are in close proximity at the nanometer scale but that are constrained from reacting until they are triggered.[11] The focus in MIC materials has changed from thin films in contact to powders. These materials, which include metals reacting with metal oxides (e.g., Al fuel with MoO_3 as an oxidizer) are of potential value for both explosives and propellants, in part because the reaction rate is tunable over several orders of velocity because it is possible to vary the intermolecular separation distance of reactants. MICs have the potential to release perhaps twice the specific energy of common explosives like TNT or HMX while providing extra margins of safety because they are insensitive to impact or electrical charge. The potential to tailor reaction rate in propellants or pressure pulses in explosives offers great benefit. MICs are the focus of current interest within DoD and DOE, but without additional resources payoff is unlikely until the midterm.

The panel was asked to identify revolutionary technologies where possible. One such area may be nuclear isomers, though it may take decades for them to be demonstrated successfully, if they can be. Nuclear isomers are nuclides that have the same mass number and atomic number but occupy different nuclear energy states (McNaught and Wilkinson, 1997). These long-lived nuclear excited states could result in high-energy, clean-gamma-ray bursts for a variety of applications; emission is "clean" because energy is released by spindown of the nucleus, with no residual radioactivity.[12] In addition to energetic materials, potential applications include satellite power, space propulsion, gamma-ray lasers, and other pulsed power uses. The specific energy is potentially 10^5 times as great (GJ/gm, with MJ output pulses) as traditional chemical fuels, yielding enormous payoffs. Much of the research required lies in the realm of nuclear physics and involves studies of x-ray-induced decay, energy release, and isomer lifetimes. However, the participation of radiochemists, materials scientists, and engineers is likely to be necessary to help transform the basic physics research into practical, self-contained DoD systems.

Logistic Fuels An overriding fuels technology concern for the military is logistics. Fewer fuels simplify the logistics burden and improve the ability

[11]Danen, W.C., Busse, J., Jorgensen, E., Smith, B., Ferris, M., Martin, J., and Dye, R., "Los Alamos Nanoenergetic MIT (Super Thermite) Program," paper presented at 221st ACS National Meeting, San Diego, CA, 2001.

[12]Carroll, J.J., "Nuclear Isomers and X-ray Driven Gamma Emission," paper presented to the Committee on Advanced Energetic Materials and Manufacturing Technologies, National Research Council, Washington, DC, October 25, 2001.

to rapidly and flexibly provide energy to the system that most requires it—ideally all military systems would run on a single fuel. Many current military propulsion systems have been designed to use a variety of petroleum-based fossil fuels, such as jet fuel (e.g., JP-5 or JP-8) and diesel. These fuel families are well-established in military applications and supported by a strong logistics infrastructure.

This panel clearly recognizes that new electrochemical propulsion and a power system employing fuel cells are in various stages of design and development, and believes development as outlined in the Energy Conversion section below is essential for DoD applications in 2020. However, this panel also believes that jet fuel and diesel fuel will continue to power most military platforms in 2020 because they have established themselves as workhorse fuels and because of the cost and time required for propulsion system redesign to enable other fuels.

Improvements to today's logistic fuel technology are expected to be largely evolutionary. An exception may be the use of microscale process intensification in which a large number of microelectromechanical systems (MEMS) devices or mesoscale processors are used in parallel to improve yield, reduce emissions, provide more rapid heat and mass transport, enhance reaction rate, and improve safety. Additional details on the rationale for these conclusions, together with discussion of potential evolutionary improvements, are provided in Appendix D.

Energy Conversion

Conversion of energy from one form to another is of paramount importance to the future of DoD. Many types of devices can be envisioned that allow this interconversion. Foremost among these is the fuel cell, which directly converts chemical into electrical energy. Because there is no combustion, the fuel cell is not limited by the Carnot cycle and therefore offers the potential for very high efficiencies. Materials challenges for fuel cells were of major interest to the panel. Other areas of energy conversion that were examined were MEMS devices for use in microturbine-based systems, and inorganic thermoelectric and photovoltaic materials. For the sake of brevity, discussion of these latter areas is limited to Appendix D, which also contains an expanded section addressing fuel cells.

The panel notes that materials for energy storage and conversion in this chapter are discussed without regard to the specific hardware configuration adopted. Such devices could ultimately appear in entirely new forms such as integration into (or with) textiles, and the form is largely a processing issue. Also, as discussed earlier and as shown in Figure 4-1, organic photovoltaic and thermoelectric materials were of interest to the Panel on Functional Organic and Hybrid Materials (see Chapter 6).

Fuel Cells

Each of the services has a vision for an all-electric future. The Army has considered electric ground vehicles whose propulsion and drives are combined with an electric or electrothermal weapon system. The Air Force and Navy have promoted visions of the all-electric aircraft and the all-electric ship (see Appendix D). Because fuel cells output electrical power directly, they will be critical to implementing this vision. In addition, fuel cells can be implemented in a distributed way, increasing the combat robustness of the power supply.

While civilian and military fuel cells face some similar development challenges (cost, lifetime, fuel flexibility), military applications place unique demands on fuel cells. For shipboard applications, the fuel cell must be highly tolerant of saline and humid environments and possess high shock resistance; in submarines, fuel cells must be equipped not only with a fuel supply but also with an oxidant supply; for large-scale power needs, military fuel cells would ideally be compatible with diesel fuels such as JP-8 and JP-3, whereas for small-scale, single-soldier applications, rapid start-up and high energy and power densities are of paramount importance. It is unlikely that any single fuel cell system will meet the needs of all military platforms, much less both civilian and military applications.

Background Fuel cells are devices that, using as their input some type of chemical fuel, output electrical power. The principle of operation of a fuel cell is illustrated in Figure 4-3 and elaborated upon in Appendix D. The fuel cells developed over the past few decades are differentiated essentially by the type of electrolyte they employ, which in turn determines the temperature at which the fuel cell can be operated. Usually, for reasons of efficiency, higher temperature operation is preferred, but for portable (intermittent) power applications, lower temperature operation is preferred.

Fuel cells offer several attractive features. In particular, (1) they exhibit very high efficiencies, easily 45 percent for high-temperature fuel cells, and even greater efficiencies in hybrid systems; (2) they offer the potential for very high energy and power densities, because they operate on hydrocarbon fuels; (3) they are easily scaled without significant loss in performance

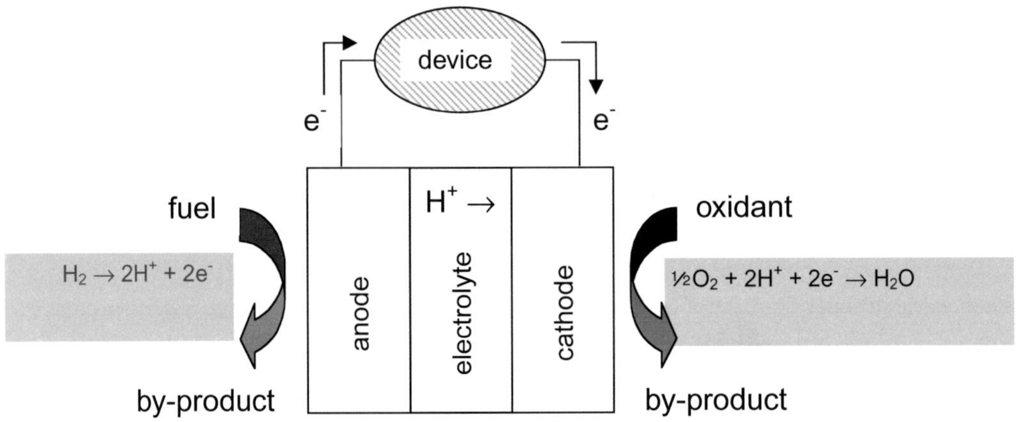

FIGURE 4-3

Schematic of a fuel cell. The overall chemical reaction is $H_2 + {}^1/_2O_2 \rightarrow H_2O$. Anode and cathode reactions given are appropriate only for proton-conducting electrolytes. The reactions would be modified for electrolytes with different mobile ions, but the general principle is unchanged.

parameters to a wide range of power demands; (4) they are nonpolluting (ultralow NO_x and SO_x emissions); (5) they can be implemented in a distributed, modular fashion, resulting in a power generation system that is robust under combat conditions; and (6) they can be operated in a closed-cycle system that uses as input solar or nuclear energy and uses fuel cell waste products to generate chemical fuels for reuse in the fuel cell (see Figure 4-4). This latter mode of operation may be particularly advantageous in future military operations, in which shipboard-based nuclear energy and high-efficiency photovoltaic devices could be used to reduce DoD reliance on increasingly scarce fossil fuels. Using electricity to produce hydrogen and oxygen that are then consumed to generate electricity in a fuel cell would reduce its round trip efficiency compared to a battery system (which could also be recharged using this approach). There are applications in which the higher energy density of a fuel cell system may be needed.

Table 4-1 provides information on types of fuel cells and gives conservative values for fuel cell operating conditions; laboratory work to extend the operating envelope to lower temperatures continues. Each of the types listed in Table 4-1 has been demonstrated in complete fuel cell systems, with alkali fuel cells (AFC) and phosphoric acid fuel cells (PAFC) being the

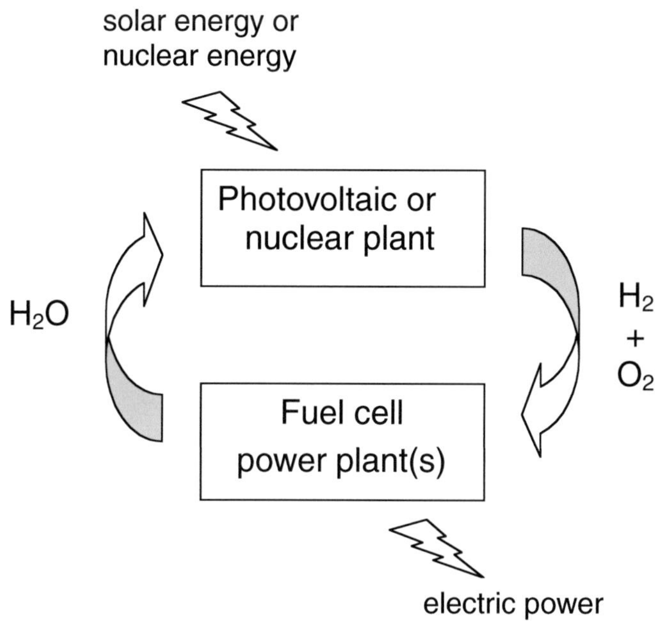

FIGURE 4-4

Fossil-fuel-independent power generation in a fuel cell.

most mature technologies, and solid oxide fuel cells (SOFC) and polymer electrolyte membrane fuel cells (PEMFC) the most recent.

Research Required to Meet DoD Needs The panel addressed the breakthroughs needed in fuel cell research to meet DoD needs within the context of three topical areas: electrolytes, electrodes and electrocatalysis, and fuels and fuel cell system design. It should be noted, however, that advances in one component or aspect will dramatically affect the others. For example, novel electrocatalysis methodologies may simplify system design. A systems approach is therefore essential. Appendix D provides a more detailed technical discussion of research opportunities in the three areas, including an enhanced discussion of ion transport.

Electrolytes A key limitation of performance for all types of fuel cells is the resistance of the electrolyte. Exploratory synthesis of new solid-state electrolyte materials, guided by computational materials science and

TABLE 4-1 Fuel Cell Types and Selected Features

Type	Temperature (°C)	Fuel	Electrolyte
Polymer electrolyte membrane fuel cell (PEMFC)	70-110	H_2, CH_3OH	Sulfonated polymers (Nafion™)
Alkali fuel cell (AFC)	100-250	H_2	Aqueous KOH
Phosphoric acid fuel cell (PAFC)	150-250	H_2	H_3PO_4
Molten carbonate fuel cell (MCFC)	500-700	Hydrocarbons, CO	$(Na,K)_2CO_3$
Solid oxide fuel cell (SOFC)	700-1000	Hydrocarbons, CO	$(Zr,Y)O_{2-\delta}$

enhanced by combinatorial approaches, is thus an area ripe for revolutionary fuel cell advances for future defense needs. Solid state conductors of either oxide ions or protons with conductivities of $\sim 10^{-3}$ S/cm at temperatures between 25 and 500°C and sufficient mechanical and chemical integrity would completely alter the fuel cell landscape.

A particular example is direct methanol fuel cells (DMFCs). Because of their relative simplicity and minimal auxiliary components, these are considered most suitable for single-soldier applications. In today's DMFCs, power densities are low because the methanol concentration in the fuel, a mixture of water and methanol, is limited to no more than ~3 percent and the electrolyte is relatively thick. Higher concentrations and thinner electrolytes lead to dangerous levels of methanol crossover (diffusion across the hydrated polymer), which can result in uncontrolled combustion. Zero methanol crossover electrolytes would reduce the amount of unnecessary water carried by the soldier or a supporting robot by a factor of about four while increasing the power density of the fuel cell by at least a factor of two. While civilian applications would also benefit from such electrolytes, the suitability of alternative fuel cell types in many of those applications (e.g., reformate-based PEMFCs) lowers the motivation of the civilian sector to address this challenge.

Electrodes and Electrocatalysis Critical to the function of a fuel cell are the electrodes/electrocatalysts. These components catalyze the electrochemical reduction of oxygen at the cathode and the oxidation of fuel at the anode. Breakthroughs in electrocatalysts would revolutionize fuel cells by giving them fuel flexibility and increasing their power outputs. However, it is not at all clear how such breakthroughs can be achieved, particularly in electrocatalysis of diesel fuels that are of unique importance to the military. The importance of electrocatalysts, combined with the absence of a clear pathway for success, dictate that the military explore a broad range of solutions to improve fuel cell catalysts.

Diesel fuel anode catalysts should be heavily emphasized because these fuels are of lesser interest to the civilian sector. Tools developed in the last few years can and should be effectively leveraged by DoD as part of this effort: (1) combinatorial chemistry methods for quickly finding and optimizing new alloy compositions; (2) computational materials chemistry for guiding exploratory research in identifying new catalysts; and (3) new synthetic methodologies that allow for architectural control of fuel cell electrodes at the nanoscale, potentially enhancing reaction kinetics by dramatically increasing electrode surface area and restricting reactions to confined regions. Though the fuel cell community has yet to take advantage of such advances, it is at the interface between these fields and traditional electrochemistry that the most significant breakthroughs can be anticipated.

Fuels and Fuel Cell System Design It is evident that the choice of fuel is a fundamental question for fuel cell systems. While higher temperature fuel cells can operate directly on hydrocarbon fuels (typically using water to achieve internal reforming), lower temperature fuel cells operate on hydrogen or sometimes methanol. Generating, storing, and transporting hydrogen is a significant hurdle. Hydrogen is typically produced from methane (CH_4) or other hydrocarbon fuels in reactors termed "reformers" that yield hydrogen tainted with trace impurities such as CO and H_2S. Removing these impurities is of paramount importance for optimal fuel cell performance.

In recent years there have emerged new approaches to generating clean hydrogen from a variety of fuel sources (including coal, which could reduce dependence on imported oil) that warrant further DoD investment. These rely on membrane reactors, which enable the production of high-purity products like hydrogen at rates beyond nominal thermodynamic limits (see Appendix D). Successful operation of membrane reactors requires materials with high hydrogen fluxes, good catalytic properties, and excellent chemical and mechanical robustness. Research in this area is in its infancy, with the majority of work directed at palladium-based membranes and some limited studies of ceramic membranes. Given the potential of such reactors to solve the daunting fuel problem for fuel cells, additional research on new membranes and nonprecious metal catalysts is warranted.

An entirely different strategy for providing hydrogen for fuel cells is the use of nonconventional fuels, such as ammonia, hydrazine, or sodium borohydrate. As with hydrogen, some other energy source must be used to synthesize these fuels. They offer the advantage, however, of being processable in a closed cycle so that reliance on fossil fuels is minimized. Particularly attractive is sodium borohydrate, which readily reacts with water to release hydrogen. The byproduct, sodium borate, can be regenerated to borohydrate using independent energy inputs.

Storage of hydrogen fuel will be of considerable importance in future military systems. Storage issues related to both conventional approaches (compressed gas, liquid hydrogen) and sorption systems have been identified (Cava et al., 2001; Schlapbach and Zuttel, 2001). Because the former depend on either high pressures or extremely low temperatures, they are problematic for the battlefield environment. Physisorption or chemisorption systems using metallic and intermetallic hydrides (based on light elements like Mg and Al) or possibly carbon-based materials may overcome these limitations and allow for high-density storage of hydrogen. Here, DoD will likely be a beneficiary of commercial work.

MEMS for Energy Conversion

MEMS extend the integrated circuit concept to incorporate electronic, electrical, mechanical, optical, magnetic, and photonic functions into a chip or structure built on a wafer of silicon or other material. An extensive manufacturing base is already available for manufacturing many MEMS devices. Indeed, most of us literally trust our lives to the accelerometer that deploys the air bags in our cars, a MEMS device that has been in mass production for years. MEMS micropower sources (gas turbines, fuel cells) and microchemical processing plants for fuel reforming are the two MEMS technologies that are most relevant to the work of the Energy and Power Panel. The Electronics and Photonics Panel report (Chapter 5) considers other MEMS applications, such as sensors.

Chip-scale microturbine engine hardware (e.g., 21 mm in diameter by 4 mm high) has already been fabricated and is under test. For DoD this technology makes possible sensor-equipped, hummingbird-sized UAVs that could be deployed at the squad level for immediate neighborhood surveillance. Table 4-2 shows comparison between the performance of

TABLE 4-2 Comparison of Initial Performance of Macro Gas Turbines and of a MEMS Microturbine

Engine	JUMO 004 (1939)	J-85 (1954)	Demo Micro (1999)
Aircraft	Me-262	T-38, F-5	MAV
Engine diameter (in.)	31	18	0.8
Thrust (lb)	2000	2950	25×10^{-4}
Thrust/wt	1.2:1	7.3:1	5.5:1
Thrust-specific fuel consumption (lb/hr/lb)	1.4	1	1.4
Turbine inlet temperature (°F)	1430	1715	2421
Overall pressure ratio	3:1	7:1	2:1

SOURCE: Waitz, I., "Micro Heat Engines," paper presented at the Grand Challenges Workshop, November 16-18, 1999.

macro gas turbines for aircraft (standard and turbofan) and a micro gas turbine, each at their initial states of development.

The thrust-to-weight ratio and thrust-specific fuel consumption (SFC) of the microturbine are remarkable for a gas turbine with a pressure ratio of 2, especially at this early state of development. Small engines are usually less efficient than larger engines; hot gas leakage is proportionately a higher percentage of total gas flow because tip clearances represent a larger fraction of the blade surface area in small engines. Yet the microengine has a thrust SFC comparable to large engines. The microengine runs uncooled at ~2400°F (1315°C). If it could be made to run hotter, it would be even more efficient, but use of silicon as the engine material limits the turbine inlet temperature (TIT) to ~90°C below the melting point of silicon. The use of alternative materials such as silicon nitride or silicon carbide could increase efficiency by raising the TIT to 1350 or 1375°C (2500 to 2550°F). However, SiC is more difficult to process than silicon. Thus it is encouraging that a SiC MEMS accelerometer for use at moderately elevated temperatures (>400°C) has recently been achieved using standard microfabrication technologies (AFRL, 2001).

MEMS microturbines can also be used for power for the individual soldier and auxiliary power sources. Wafer-level integration of multiple microturbines may enable modular power sources in increments of 40 to 50 W (the power level of one microturbine). Using today's technology, 100 microturbines could be fabricated within one 6-layer stack of silicon wafers. By 2020, significant size reductions are likely, as semiconductor fabrication technology continues to reduce feature sizes (currently 0.18 down to 0.13 microns). New thermal managment and heat disipation technologies would promote the longevity and efficiency of these devices and will therefore be critical for their use in the field.

MEMS technology is also being developed for 10- to 500-milliwatt fuel cells and reformers.[13] Wafer-level integration of such fuel cells and reformers (both of which could be fabricated on the same wafer) would enable a scaling up of power in increments determined by the replicated devices. The materials issue for this technology is to find a suite of anode, cathode, interconnect, and ion exchange membrane materials that can withstand the highly reactive environments encountered in semiconductor fabrication processing. Other micropower systems, such as chip-level combustors and thermophotovoltaic systems, are also possible. The energy densities of diesel fuel and/or methanol are each at least ten times greater than lithium-ion batteries.[14] However, to realize this advantage the fuel processors and ancillary equipment need to be simplified and miniaturized. The panel believes that MEMS-based, subwatt microfuel cells may enable the individual soldier to move beyond the limitations inherent in battery-powered communications and sensor systems.

To move advanced MEMS approaches into DoD applications for reliable and durable use in energy and power applications (particularly in dynamic as opposed to static applications), other materials and process issues must be considered. Extension to nanoelectromechanical systems (NEMS) will only add to the difficulties. Therefore, it is essential to understand the phenomena occurring at the micro- and nanoscales, and the differences from conventional phenomena due to scaling factors. Key areas for research are (1) tribology; (2) heat transfer thermal stresses and mechanical properties; (3) corrosion and other environmental degradation modes; and (4) reactions between dissimilar materials at elevated tempera-

[13]Franz, A., "Micro-Reformers," paper presented at Grand Challenges Workshop, MIT, Boston, MA, Nov. 16-18, 1999; Tang, W., "MEMS Program at DARPA: Past, Current, and Future," paper presented to the Panel on Energy and Power Materials of the Committee on Materials Research for Defense After Next, National Research Council, Irvine, CA, October 11, 2001.

[14]Jones, E., "Sub-Watt Power Using an Integrated Fuel Processor and Fuel Cell," paper presented at IMRET 5 2001, International Conference on Microreaction Technology, Strasbourg, France, May 2001.

tures. All of these must be understood at the micro (0.5- to 5-micron) and nanoscales (10- to 500-nm). Of special importance are differences in physical properties at the nanoscale, where a large fraction of atoms may be in surface rather than volume lattice sites. The panel believes that MEMS for power and energy is an area full of military promise, but it will require significant materials science and engineering support to reach its full potential.

Small Nuclear Power Sources

Small nuclear-power devices are typically implemented as radioisotopic thermoelectric generators (RTGs), beta cells, and similar devices. They rely on the nuclear reaction to produce heat that is converted to electricity by thermoelectrics (RTGs) or direct conversion of radioactive decay particles (beta cells). Output power is a few milliwatts to approximately 1 kW; applications are typically in spacecraft or weapon systems. While the panel did not have sufficient resources to explore this class of nuclear power sources, it did review earlier studies (NRC, 1997, 1999) and other information that suggests that this area should be explored. In particular, a hybrid device made of a small nuclear power source having essentially a constant power output combined with a battery or electrochemical capacitor could be advantageous. While concerns about contamination, toxicity, and environmental acceptability must be addressed, the panel believes that such a device could offer the possibility of a practical, acceptable power source for low-power electronics.

Electrical Power Generation and Transmission for Propulsion and Related Systems

Platform propulsion and support using more electric systems are of central importance to the Army, Navy, and Air Force of 2020. Although the panel did not have sufficient resources to fully identify all the materials challenges in this important area, it identified factors affecting the consumption of, and the benefits to be derived from, increased electric power use for platform propulsion and support over the next 20 years. These factors, which have a multiplier effect, are detailed in Appendix D. Among them are identification of unique DoD benefits as well as discussion of characteristics of materials (e.g., high-temperature superconductors) that will be required to make all-electric/more electric systems a reality.

Kinetic Energy Dissipation and Protection

This section briefly summarizes advanced lightweight armor materials (e.g., body armor) and transparent materials for dissipating kinetic energy. Other areas of energy dissipation (laser, structural armor, etc.) are discussed elsewhere in this report.

The most important asset and most complex system on the battlefield is the individual soldier. Protecting soldiers from fragments and small-arms kinetic energy (KE) threats is perhaps the single most important energy dissipation issue for DoD. With the trend toward a smaller but more capable force, even a relatively small number of casualties may severely degrade mission effectiveness. Because major improvements in armor materials and designs may be classified, highly promising existing and developmental materials and designs may have been omitted in this study, which is limited to unclassified sources.

Lightweight Armor

Three classes of nonstructural armor were considered by the panel: soft (fabric) personnel armor capable of defeating fragments and low KE ball ammunition (presently typified by Kevlar fabric); hard-faced composite light personnel armor for defeating high energy ball and armor piercing (AP) rounds (presently typified by ceramic tiles backed by layers of Kevlar); and transparent armor for face, riot shields, and lightweight vehicle windshields (currently typified by ballistic glass/polymer multilayers).

These engineered systems require trade-offs among such factors as penetration resistance, weight, bulk, deflection, multihit capability, flexibility, comfort, and field durability. Although this deals with materials properties, ballistic performance is in fact a systems property. Projected increases in the ballistic resistance of materials, while important, are unlikely in and of themselves to make dramatic increases (i.e., >100 percent) in ballistic efficiency for all classes of personnel armor. Rather, such materials performance increases will enable improved or innovative designs, which will synergistically leverage the materials property gains to yield dramatic armor systems improvements. The next section summarizes principal conclusions. Potential lightweight armor goals between now and 2020 are discussed in Appendix D.

Soft Fabric—Fragmentation Protection Of the classes of armor considered in this assessment, soft, textile-fiber-based armor for fragment and low-velocity ball rounds is the best understood from a theoretical basis. Cunniff (1999a) found that, given projected improvements in fiber properties, reductions in areal densities of at least 50 percent for a given threat level are possible by 2020. This increased performance may entail excessive behind-armor deflection, necessitating redesign of the entire soldier protection system to compensate. Introduction of new fibers could provide an impetus if a clear "requirements pull" should justify the cost of new fiber development. Some fiber families (e.g., PIPD) employ rigid rods within the chain combined with strong interchain hydrogen bonding. Continuous fibers based on single-walled carbon nanotubes could be potentially significant if processing can control structure and limit defects. Hybrid yarns employing fibers having complementary characteristics could also provide synergy.

Hard-Faced Protection Against Ball and AP Rounds While the conventional wisdom in hard-faced armor systems is that the thickness of the hard face must be near that of the projectile diameter, recent work on the defeat of ball ammunition has challenged this rule of thumb. Cunniff (1999b) has shown that judicious design can achieve equal protection with less damage by using thinner face plates of the same ceramic. There may also be benefits derivable from engineered nanomaterials for protection against such threats. Withers has tested nanomaterial configurations against ball projectiles and shown potential benefits.[15] The wide range of possible nanomaterials and configurations makes this a fertile ground for examination; the challenge is to narrow the range appropriately in order to limit the testing required. Computational modeling could enhance understanding of the behavior of such materials, but a real payoff could be realized if an accurate, small-scale lab test could be developed to predict and supplement results of full-scale configuration testing.

Ceramic-faced armor for personnel protection against AP projectiles dates from the early 1960s with the development of boron carbide/GRP lightweight composite armor (Viechnicki et al., 1991). In general, the material must be harder than the projectile and should have as high an impedance ($I = \sqrt{\rho E}$) as possible. The most efficient hard-face materials tend to be the oxides, carbides, or nitrides of Al, Si, or B. Currently, B_4C

[15]Withers, J.C., chief executive officer, Mer Corp., Tucson, AZ, private communication, September 26, 2001.

provides the most mass-efficient hard-face material with a density of ~2.5 g/cc. There are presently no known materials that have significantly higher impedance combined with a lower density; the properties of some armor ceramics are shown in Table 4-3.

In the panel's view, only incremental improvement (<30-40 percent), can be expected in the ballistic resistance of rigid and hard-faced armor systems based on straight materials improvements. Similarly, incremental improvements in design are not likely to yield performance increments of the order of 100 percent. What is required are totally new approaches in armor design that lead to new mechanisms for defeating the more severe small arms threats, such as 0.30 cal AP. Possibilities include projectile tipping or rotating, momentum trapping, and confinement.

Transparent Protection Currently, there are three main developmental transparent armor ceramics: AlON (aluminum oxynitride), spinel (magnesium aluminate), and sapphire (single-crystal aluminum oxide). Sapphire is currently the most industrially available of these materials, and AlON appears to be the most ballistically promising. However, all three materials have a density in the range of 3.6 to 4 g/cc. Even though these materials perform significantly better than glass/polycarbonate laminates, they still have areal densities nearly double those of ceramic-faced opaque armors. It is therefore critical to try to synthesize new transparent ceramics that have densities close to that of boron carbide (2.5 g/cc).

Armor Materials Research Needs: Summary Progress in fibers will lead to sizable improvements in soft armor only if DoD takes the initiative to fund new fiber development. Improvements in ballistic performance of ceramics for hard-facing of lightweight armor will most likely be incre-

TABLE 4-3 Properties of Armor Ceramics

Material	Density (g/cc)	Hardness (GPa)	Young's Modulus (GPa)
AlN	3.2	14	280
Al_2O_3 (90 percent)	3.6	20	275
B_4C	2.5	30	445
SiC (sintered)	3.2	27.5	390

mental (20 to 40 percent range). For opaque lightweight armor, the most promising approach to attain major gains in performance (~100 percent) may be by innovative design. Significant increases in transparent armor performance can be made by implementing *systems* based on the current generation of developmental materials (i.e., AlON and sapphire), but further advances in the performance of ceramic-faced transparent armors will require new transparent ceramics with a density of less than 3 g/cc (with other properties equal to or better than AlON). Finally, the panel emphasizes that nanomaterials technology should be explored as a potential "paradigm breaker" in light-armor technology.

DoD Reliance on Energy Sources

National energy dependence, particularly with respect to reliance on energy imports, has been the subject of considerable policy discussion for many years. Though this is not a materials development issue per se, the committee agreed that a brief discussion of overall DoD energy usage, dependence on various sources, and issues related to DoD energy dependence through 2020 was desirable. The committee's conclusion was that, although new approaches to energy harvesting (e.g., photovoltaics) will be combined with increasing use of fuel cells, existing infrastructure makes it likely that current fuels (e.g., JP-8, diesel) and nuclear power (for the Navy) will still make up the bulk of energy sources used by DoD platforms of 2020. Supporting information is given in Appendix D.

RESEARCH AND DEVELOPMENT PRIORITIES

The energy and power area spans numerous applications, device types, materials families, and subspecialties of expertise. Nevertheless, the materials R&D priorities that would most benefit future DoD systems can be articulated in a short list, given below.

Nanomaterials Science and Engineering

A common theme throughout the report of the Panel on Energy and Power Materials is the potential contribution of materials either containing nanoscale components or having structure control at the nanometer scale.

Materials exhibit wholly different properties at the nanoscale, which often translate into radically different macroscopic behavior. Nanostructuring offers a powerful lens through which to both reexamine the vast array of existing materials and to approach new materials.

Nanomaterials science and engineering affect virtually every energy and power area. For batteries, nanocomposite electrolytes could lead to entirely new mechanisms of ionic transport with possibility for very high conductivities. In dielectric capacitors, copolymers with structured nanodomains have the potential that may greatly increase dielectric constants. In energetic materials, nanoscale aluminum powders are already used to tailor burn rate in propellants; engineering of energetic material nanocomposites by controlling such factors as nanoparticle size and size distribution, composition, and morphology may lead to major improvements in these materials. New synthetic methodologies that allow for architectural control of fuel cell electrodes at the nanoscale could enhance reaction kinetics by dramatically increasing electrode surface area and restricting reactions to confined regions.

The use of nanoscale and nanostructured materials spans the entire range of metals, ceramics, polymers, and composites of interest to DoD, and the paybacks are potentially huge. The challenge is to produce nanostructures that have useful material properties in a controlled and reliable manner, while assuring that the structures remain stable over extended use. Accordingly, DoD should continue to support efforts to develop new ways to structure materials and to combine them with other materials at that nanoscale.

Engineered Interfaces and Surfaces in Materials

The importance of engineered interfaces and surfaces in materials arose continually, either directly or indirectly, as the Energy and Power Materials Panel identified materials challenges. Electrochemical capacitors will have improved power density if the interfaces on which charge is stored are tailored to provide electrochemical stability as well as optimum charge storage. Similarly, lithium batteries with interfaces tailored to eliminate pyrophoric dendrites could lead to 2-3 times power density improvements. Advanced approaches for increasing the surface area of intermixed components in energetic materials may be able to significantly raise the energy/mass ratio, leading to smaller projectiles of similar or greater capability. MEMS devices may be dramatically improved by an enhanced understanding of reactions occurring at surfaces between dissimilar materials at elevated temperatures. DoD should place renewed emphasis on interfaces and surfaces, either separately or under a nanomaterials initiative.

Advanced Energy Storage and Conversion Materials

New materials will be essential in meeting the energy and power needs of DoD in 2020. The largest subarea that can potentially benefit is materials for electrochemical energy storage and conversion for use in batteries, electrochemical capacitors, fuel cells, and fuel reforming. As an example, high-surface-area battery electrodes made using a one-step process may allow full exploitation of the material's possibilities. Synthesis of new solid-state electrolytes (e.g., perovskite or fluorite oxide conductor) could produce revolutionary advances for the DoD of 2020. For instance, a zero methanol crossover electrolyte would reduce the water requirements for a direct methanol fuel cell by a factor of four while at least doubling the power density, thus greatly reducing the burden of water carried by a dismounted soldier or supporting robot. New electrocatalysts for reforming diesel fuel could have a huge DoD payoff. Similarly, diesel fuel anode catalysts should receive considerable attention, because the civilian sector has little interest in this area, while it is crucial to DoD for reforming logistic fuels.

The panel also suggests that novel energetic materials, including propellants and explosives, continue to be a fruitful area for research for future defense needs. Some of the key subareas of this important category have been highlighted in this chapter.

Tools for Accelerated, Systematic Materials Discovery

Considerable time may separate the discovery and the introduction of a new material, depending on the application area and the forces driving the new material. Structural materials have required on average approximately 20 years for introduction; energetic materials have often taken considerably longer. Traditionally, the development of new materials has relied on the intuition, ingenuity, and knowledge base of experienced individuals who can extrapolate the properties and structures of existing materials to those of new materials. To be exploited, materials must first be discovered.

To speed the discovery of new materials, the panel emphasizes, tools for accelerated systematic materials discovery and application should be employed to the maximum extent possible. These tools include (1) computational materials science, to assist in analytical design of promising new materials; and (2) combinatorial materials science, to speed the experimental discovery of new materials. The latter is particularly important in those cases where there are few signposts to promising materials, such as fuel cell electrocatalysts, or where it appears desirable to map out possible limits on the material properties of a family of compounds. DoD is already using these tools effectively. This suggestion emphasizes the importance of expanding their use in materials discovery and application, to hasten the discovery and exploitation of novel materials.

Materials as the Foundation for Systems

Materials are the basic ingredients of all DoD hardware systems. They (together with the design) affect system cost throughout the system life cycle in terms of the costs for raw materials, processing and fabrication, assembly, NDI/NDE, repair and maintenance, and disposal. It is axiomatic that much of the cost associated with a system is locked in very early in system development by the materials and design decisions made at that time. It is also clear that many materials scientists often become so strongly associated with a material or family of materials that they become advocates for that material. Similarly, materials and design engineers often become wedded to particular approaches.

During the investigations conducted by the Energy and Power Panel, it was not uncommon to encounter strong advocates for particular materials. Typically, one or more key properties of a material, which happened to be optimized and which made the material a candidate for an application, were cited as evidence for the material's value in that application. In reality, however, every material has strengths and weaknesses. Virtually every application places multiple requirements upon a given material, making the selection of material a complex tradeoff among different, often conflicting, factors.

Similarly, the cost savings for using a given material in a component were sometimes provided as a "point comparison" with another material, without regard to the contribution of that component to total system cost. Such comparisons provide little real insight. For example, a system (e.g., a fuel cell) is a collection of interdependent components (e.g., anode, cathode, electrolyte), and the introduction of one material into a system has consequences for every other material used in that system in terms of such issues as reactivity, thermal stability, or need for compatibility coatings.

Systems can be quite complex. They require a broad view, while materials research requires that the materials scientist focus largely on one or only a few aspects of material structure or behavior. Therefore, it is only natural that advocacy occurs. However, DoD materials science and engineering managers should emphasize a systems-based approach in which all material properties relevant to a class of applications are identified as early as possible, and in which point solutions are avoided. In addition to challenging point solution advocacy wherever it occurs, this would entail evaluating materials in a subscale device (or minisystem configuration) as early as possible in the research stage. This should help identify material strengths and weaknesses and should enable decisions to be made as early as possible about the potential effectiveness or limitations of a material in a class of applications.

REFERENCES

AFRL (Air Force Research Laboratory). 2001. Scientists are developing next-generation microelectromechanical systems from silicon carbide. P. 29 in AFRL Technology Horizons. Ohio: AFRL, Wright Patterson Air Force Base.

Bates, J.B. 2000. Thin film lithium and Li ion batteries. Solid State Ionics 135: 33.

Bronstein, L.M., C. Joo, R. Karlinsey, A. Ryder, and J.W. Zwanziger. 2001. Nanostructured inorganic-organic composites as a basis for solid polymer electrolytes with enhanced properties. Chemistry of Materials 13(10):3678-3684.

Cava, R., F. DiSalvo, L. Brus, K. Dunbar, C. Gorman, S. Haile, L. Interrante, J. Musfeldt, A. Navrotsky, R. Nuzzo, W. Picket, A. Stacey, and A. Wilkinson. 2001. Future Directions in Solid State Chemistry, Report of NSF-Sponsored Workshop, October, UC-Davis. Arlington, VA: National Science Foundation.

Christen, T., and M.W. Carlen. 2000. Theory of ragone plots, J. Power Sources 91:210-216.

Clelland, I., R. Price, and J. Sarjeant. 2000. Advances in capacitor technology for modern power electronics. Pp. 145-148 in Proceedings of the 24th International Power Modulator Symposium, June, Norfolk, VA. Piscataway, NJ: IEEE.

Croce, F., G.B. Appetecchi, L. Persi, and B. Scrosati. 1998. Nanocomposite polymer electrolyte for lithium batteries. Nature 394(6692):456.

Cunniff, P.M. 1999a. Dimensionless parameters for optimization of textile-based body armor systems. Proceedings of the 18th International Symposium on Ballistics, November, San Antonio, TX. Lancaster, PA: Technomic.

Cunniff, P.M. 1999b. Assessment of small arms (ball round) body armor systems. Proceedings of the 18th International Symposium on Ballistics, November, San Antonio, TX. Lancaster, PA: Technomic.

DoD (Department of Defense). 1999. Militarily Critical Technologies (MCT) Part I: Weapons System Technologies. Fort Belvoir, VA: Defense Technical Information Center.

Dong, W., D.R. Rolison, and B. Dunn. 2000. Electrochemical properties of high surface area vanadium oxide aerogels. Electrochemical and Solid State Letters 3(10):457.

Doughty D.H. 1996. Materials development for lithium-ion batteries. Pp. 102-107 in Electric and Hybrid Vehicle Technology '96: The International Review of Electric and Hybrid Vehicle Design and Development, G. Lindsay, ed. Dorking, Surrey, UK: UK & International Press.

Hamlen, R.P. 1995. Metal/Air batteries. Pp. 38-1 to 38-45 in Handbook of Batteries, 2nd Edition, David Linden, ed. New York: McGraw-Hill, Inc.

Kornmann, X., L.A. Berglund, J. Sterte, and E.P. Giannelis. 1998. Nanocomposites based on montmorillonite and unsaturated polyester. Polymer Engineering and Science 38(8):1351-1358.

McNaught, A.D., and A. Wilkinson. 1997. IUPAC (International Union of Pure and Applied Chemistry) Compendium of Chemical Terminology, 2nd Edition. P. 54:1545. Oxford, UK: Blackwell Science.

Messersmith, Phillip B., and Emmanuel P. Giannellis. 1995. Synthesis and barrier properties of poly(e-caprolactone)-layered silicate nanocomposites. Journal of Polymer Science: Part A: Polymer Chemistry 33:1047-1057.

National Research Council. 1997. Energy Efficient Technologies for the Dismounted Soldier. Washington, DC: National Academy Press. Available online at <www.nap.edu/catalog/5905.html>. Accessed October 15, 2001.

National Research Council. 1999. Reducing the Logistics Burden for the Army After Next: Doing More with Less. Washington, DC: National Academy Press. Available online at <www.nap.edu/catalog/6402.html>. Accessed January 15, 2002.

Pang, S.C., and M.A. Anderson. 2000. Novel electrode materials for ultracapacitors: Structure and electrochemical properties of sol-gel derived manganese dioxide thin films. P. 415 in Proceedings of New Materials for Batteries and Fuel Cells, San Francisco, CA, D.H. Doughty, L.F. Nazar, M. Arakawa, H.P. Brack, and K. Naoi, eds. Warrendale, PA: Materials Research Society.

Poizot, P., S. Laruelle, S. Grugeon, L. Dupont, B. Beaudoin, and J.M. Tarascon. 2000. Electrochemical Reactivity and Reversibility of Cobalt Oxides Towards Lithium. Comptes Rendus de l'Académie des Sciences, Series II, p. 681. Paris: Elsevier.

Potomac Institute. 2000. Conference Summary Report of Out of the Box and Into the Future: A Dialogue Between Warfighters and Scientists on Far-Future Warfare. Arlington, VA: Potomac Institute. Available online at <http://www.potomacinstitute.org/pubs/otb_summary.pdf>. Accessed December 15, 2001.

Riley, M.W., P.S. Fedkiw, and S.A. Khan. 2000. Nanocomposite based electrolyte for lithium-ion batteries. P. 137 in Proceedings of New Materials for Batteries and Fuel Cells, San Francisco, CA, D.H. Doughty, L.F. Nazar, M. Arakawa, H.P. Brack, and K. Naoi, eds. Warrendale, PA: Materials Research Society.

Salkind, A.J. 1998. Advances in battery technologies and markets—Materials science aspects. P. 1 in Proceedings of Materials for Electrochemical Energy Storage and Conversion II—Batteries, Capacitors and Fuel Cells, Boston, MA, D.S. Ginley, D.H. Doughty, B. Scrosati, T. Takamura, and Z.M.J. Zhang, eds. Warrendale, PA: Materials Research Society.

Sata, N., K. Eberman, K. Eberl, and J. Maier. 2000. Mesoscopic fast ion conduction in nanometre-scale planar heterostructures. Nature 408(6815):946.

Schlapbach, L., and A. Zuttel. 2001. Hydrogen-storage materials for mobile applications. Nature 414(6861):353.

Singh, D., R. Houriet, R. Vacassy, H. Hofmann, V. Craciun, and R.K. Singh. 2000. Pulsed laser deposition and characterization of $LiMn_2O_4$ thin films for application in Li ion rechargeable battery systems. P. 83 in Proceedings of New Materials for Batteries and Fuel Cells, San Francisco, CA, D.H. Doughty, L.F. Nazar, M. Arakawa, H.P. Brack, and K. Naoi, eds. Warrendale, PA: Materials Research Society.

Viechnicki, D.J., M.J. Slavin, and M.I. Kliman. 1991. Development and current status of armor ceramics. Ceramic Bulletin 70(6):1035-1039.

CHAPTER FIVE

Electronic and Photonic Materials

CHAPTER SUMMARY

The Panel on Electronic and Photonic Materials was charged with examining materials research needs for future defense systems in electronics, optoelectronics and photonics, and microsystems (including sensor systems). The tremendous vitality and innovation of the private sector allowed the panel to consider which future defense needs could be met by making use of commercial developments in industry and which were so specific to DoD as to require DoD investment.

The panel began by examining the following military needs that would particularly benefit from advances in electronic and photonic materials:

- Detection, identification, and defense against or avoidance of threats;
- High-fidelity imaging signals;
- Communications systems;
- Compact systems to transmit at very high power and high frequency;
- Enemy identification and monitoring;
- Dynamic camouflage/stealth; and
- Health monitoring of equipment and personnel.

While this panel considered a wide variety of military needs from several vantage points, ranging from individual devices or components to entire miniature systems, a number of common themes emerged that point to important areas for research:

- Improvements in the fundamental understanding of existing materials that have been identified as promising for military applications;
- The search for new materials with extreme properties, sometimes orders of magnitude better than what is currently available;
- New ways of combining materials, particularly at the nanoscale, in order to obtain new functionality, with particular attention to ways to routinely combine inorganic and organic or even biological materials;
- Consideration of packaging and thermal management issues, in parallel with research on materials for the devices themselves;
- Materials processing to effect the successful introduction of new materials and the optimization and manufacturability of existing materials (low-temperature processes are generally desirable and sometimes essential, particularly as device dimensions shrink and as increasingly dissimilar materials are used in close proximity to one another); and
- Theory and modeling as indispensable partners of experimental efforts in the search for and optimization of new materials, especially as nanoscale structure and control become paramount.

INTRODUCTION

The Panel on Electronic and Photonic Materials was charged with examining materials research needs for future DoD systems in three areas: (1) electronics; (2) optoelectronics and photonics; and (3) microsystems, including sensor systems. Progress in electronic and photonic materials and systems that are heavily dependent on them, such as microsystems and sensors, is occurring at an astonishing rate, fueled in large measure by tremendous investments by the private sector. This panel considered which future defense needs could be met by taking advantage of commercial developments and which were so specific to DoD as to require DoD investment. Each section of this chapter discusses (1) commercial drivers; (2) DoD drivers where materials have potentially high impact; and (3) priorities for DoD-funded materials research.

DOD NEEDS FOR ELECTRONIC AND PHOTONIC MATERIALS

The panel considered the following military needs, identified by DoD leaders,[1] to be particularly ripe for resolution via advances in electronic and photonic materials:

- *Detection, identification, and defense against—or avoidance of—threats.* Low-cost, highly capable commercial technologies are increasingly enabling even nations with very limited resources to mount considerable regional threats based on precision strikes from their own territories. This development, coupled with the periodic U.S. need to mount operations in a wide variety of conditions, requires increased reliance on systems that can detect, identify, defend against, or avoid many types of threats. Because such systems must be deployed on military equipment, as distributed sensors, or on individual soldiers, they must be compact, light, energy-efficient systems with broad capabilities.
- *High-fidelity imaging signals.* Imaging systems that could be used to identify camouflaged soldiers and materiel by soldiers or on vehicles (manned or unmanned) would reduce casualties and materiel loss.
- *Compact systems to transmit at very high power and high frequency.* Compact high-power systems would allow operation of communications and surveillance equipment at greater distances. Size and weight will determine whether such systems are appropriate for use on unmanned vehicles or by individual soldiers. Operation in new high-frequency regimes allows for higher bandwidth as well as access to spectroscopic information that is not accessible in other frequency ranges.
- *Communications systems.* Increased coordination among forces, the need for global awareness, and increased reliance on information from remote sensors and unmanned vehicles all drive the need for compact, energy-efficient communications systems that can transmit and receive securely at high bandwidth.

[1]Andrews, M., "Army Vision and S&T: Accelerating the Pace of Transformation," briefing presented to the Committee on Materials Research for Defense After Next, National Research Council, Washington, DC, February 15, 2000. Delaney, L., "Air Force Modernization," briefing presented to the Committee on Materials Research for Defense After Next, National Research Council, Washington, DC, February 15, 2000. Harwell, K., "Air Force Research Laboratory: Technology Vision," briefing presented to the Committee on Materials Research for Defense After Next, National Research Council, Washington, DC, February 16, 2000. Henley, L., "The Revolution in Military Affairs After Next," briefing presented to the Committee on Materials Research for Defense After Next, National Research Council, Washington, DC, February 15, 2000. Marshall, A., "Overview of DoD Vision and System Needs," briefing presented to the Committee on Materials Research for Defense After Next, National Research Council, Washington, DC, February 15, 2000. Marshall, A., "Overview of DoD Vision and System Needs," briefing presented to the Committee on Materials Research for Defense After Next, National Research Council, Washington, DC, January 29, 2002. Vickers, M., "The Revolution in Military Affairs (RMA)," briefing presented to the Committee on Materials Research for Defense After Next, National Research Council, Washington, DC, February 15, 2000.

- *Enemy identification and monitoring.* Monitoring of enemy troop and materiel movement, as well as discrimination between enemy and friendly elements, will become increasingly important as future conflicts move to new battlefields, such as urban environments.
- *Dynamic camouflage/stealth.* The ability to dynamically change the spectroscopic signature of surfaces, either on soldiers or equipment, would greatly reduce the probability of detection by the enemy.
- *Health monitoring of equipment and personnel.* Monitoring of personnel for health or exposure to chemical or biological agents will become increasingly important. Monitoring the health of equipment in the field is also a longstanding need, as already detailed in Chapter 3.

SPECIFIC AREAS OF OPPORTUNITY

Below, DoD needs in three specific areas of opportunity are discussed: electronics, optoelectronics and photonics, and microsystems.

Electronics

Commercial Drivers

The silicon-based microelectronics industry has been a tremendous force in the U.S. economy for at least two decades. The phenomenal ability of the industry to keep pace with, and even exceed, Moore's Law (doubling the logic density every 18 months, with a concomitant reduction in cost/bit) has put unprecedented data processing and storage in the hands of every American. The trend is expected to continue for another decade, as documented in the 2001 edition of the *International Technology Roadmap for Semiconductors* (ITRS) (SIA, 2001). DoD can leverage this revolution by taking advantage of industrial developments. For the longer term, the conclusion of the ITRS is that "10-15 years in the future, it becomes evident that most of the known technological capabilities will approach or have reached their limits." It projects that a more cost-effective alternative to planar silicon complementary metal oxide semiconductor (CMOS) technology will be required in this time frame. The solution may involve new materials, new cell architectures, and new processes.

This forecast represents the first projection that scalable planar silicon CMOS is reaching a limit. Along with this singular discontinuity several other critical material needs are identified in this roadmap for the commercial semiconductor segment over 10-15 years:

- Mask materials and photoresists for <45-nm lithography,
- Wafer materials and crystal-growing technology for >300-mm wafers (450 mm), and
- Yield model development for new materials and integration.

DoD needs to support research into novel materials for nanoscale integration beyond the limits of CMOS scaling. The commercial sector is driving the increase in information storage capability at the same rate as Moore's Law, although the relatively high level of DoD investment has enabled this industry to take advantage of new materials developments more aggressively than has happened for semiconductors.

DoD Drivers with Potentially High Materials Impact

During this century U.S. effectiveness in battle will depend on the superiority of our ability to gather information that allows us to identify, locate, and destroy or at least avoid the threat (see Figure 5-1). In spite of very rapid progress in electronic materials and systems, it has not been possible to keep up with the increasing demand for compactness, power (high or low, depending on the application), simplicity in packaging, and signal purity. Since many future military systems will be space based, use lighter platforms, or be deployed on individual soldiers, these demands are becoming imperatives. Significant improvement over existing capabilities can be achieved only by inserting novel high-performance electronic materials into sensors, detectors, and components.

The following military technologies are particularly ripe for improvement:

- Detection, identification, and defense against—or avoidance of—threats,
- Compact systems to transmit and receive very high or very low power at very high frequencies, and
- Embedded technologies that enable these systems.

Certain components in these systems are in particular need of materials advances, including

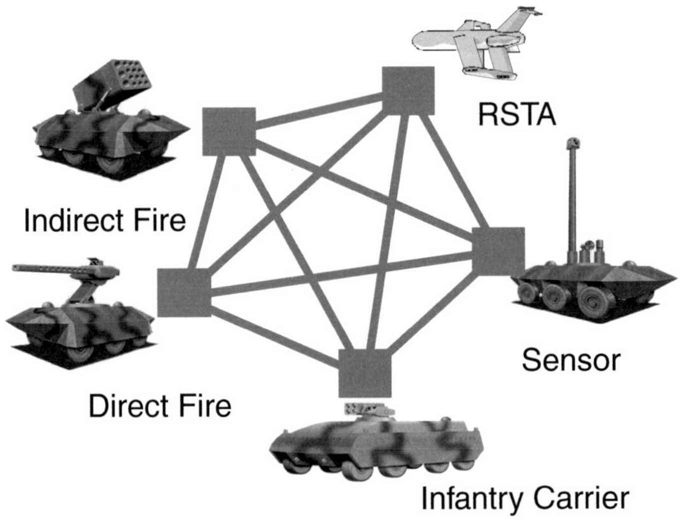

FIGURE 5-1

Relationships in the Future Combat System. SOURCE: Browning, V.M., "An Overview of Electronic Materials," briefing presented to the Panel on Electronic and Photonic Materials of the Committee on Materials Research for Defense After Next, National Research Council, Washington, DC, March 29, 2001. RSTA = Reconnaissance, Surveillance, and Targeting Antenna.

- Transmitters, receivers, and detectors for high-frequency broadband systems;
- High-power electronic components;
- Analog/digital converters;
- Electronic beam steering;
- Radio frequency (RF) microsystem components;
- Acoustic components; and
- Ferroelectric materials.

In addition, materials developments that are on the horizon or that may be envisioned could provide the basis for revolutionary technological advances; among them are superconducting materials and nanostructured materials.

Transmitters, Receivers, and Detectors for High-Frequency Broadband Systems The military needs materials for transmitters, receivers, and detectors that can operate at frequencies from MHz to over 100 GHz frequencies. SiGe is seen as the next-generation material for high-speed, high-frequency applications, but it is not suitable at the highest frequencies; these require different materials, most likely the group III-nitride crystals and their derivatives. Materials research should concern itself with developing appropriate substrates, optimizing epitaxial growth (which will require better understanding of fundamental aspects of the growth of these polar materials), and improving the junctions and contact layers (which will entail identifying important defects and ways to eliminate or control them).

High-Power Electronic Components The high-power transmitters and receivers in future systems cannot be made using traditional semiconductors like Si or GaAs. Materials for low-frequency and high-power applications like SiC and its derivatives are promising, but achieving the required signal purity must be rooted in a fundamental understanding of how to control their chemistry, crystal growth, epitaxy, and defect density.[2] The group III-nitride materials are also promising, but meeting the operating requirements for high-power applications or in hostile ambients will require interconnects and overlayers that can function around 500°C. In addition to detailed understanding of crystal growth and the control of important defects in the materials, attention must be paid to high-temperature contact materials, overlayers, and joining materials. A hybrid of carbides and nitrides of gallium, aluminum, and boron is one promising approach, although the monumental challenges of design, fabrication, and processing are still to be solved.

Analog/Digital Converters Electronic analog-to-digital converters (ADC) have continued to show steady but slow progress in resolution and speed (Walden, 1999). Performance limiters due to sampling aperture jitter and comparator ambiguity are directly related to device speed; electronic materials that yield faster devices will accelerate progress. Optical ADC and optically assisted ADC have shown promise for improving perfor-

[2]Petroski, K.J., "High Power SiC Microwave Devices for Amplifier Devices," paper presented at GOMAC, San Antonio, TX, 2001.

mance but they need new materials with 500 times higher electro-optical coefficients than lithium niobate or other oxide materials. Preliminary experiments suggest that semiconductor materials like rubidium hydrogen selenate class materials, solid solutions, or substituted II-VI, II-VI materials have potential (Guilbert et al., 1998).

Electronic Beam Steering One of the biggest challenges facing defensive systems is replacement of the mechanical or optical (mirror) beam steering with all-electronic laser-beam steering. This could be achieved with materials whose damage threshold, modulation capabilities, and transparency properties are amenable to electronic steering. Tailoring the Bragg diffraction of materials is one promising approach. As a function of microwave frequency, the angle, power, and wavelength for steering could be controlled if a material with suitable diffraction characteristics could be produced.

RF Microsystem Components Microelectromechanical microwave switches have a wide range of applications in military systems. They have extremely low insertion loss (0.2 dB) and crosstalk over a broad microwave frequency range (>40 GHz). Microwave phase shifters based on MEMS RF switches have shown much lower insertion loss than conventional electronic phase shifters. This reduces the number of amplifiers needed and also the size, weight, and power of phased-array antenna systems. A MEMS RF switch could also be used in frequency agile filters. Materials affect the reliability of these switches. Engineering of metal electrodes with low contact resistance and no stiction during separation is critical for metal-to-metal contact switches, especially for hot switching under high-power conditions. Prevention of dielectric charge-up is essential for capacitive switches. RF MEMS components will also play an important role in future wireless communication systems, as discussed below.

Acoustic Components Improved materials are needed for space and underwater acoustics. The materials properties that are important are power capability, sensitivity, noise performance, size, weight, robustness, shock resilience, aging, manufacturability, cost, and bandwidth. Materials must meet stringent performance criteria, including high Q, low loss, and temperature-stable operation. Artificially structured ternary and quaternary materials are promising. Theory and modeling will be important in finding material solutions to the severe constraints of this problem.

Ferroelectric Materials Ferroelectric materials have high dielectric constant, wide tunability, or both. Unfortunately, current ferroelectrics are limited to low dielectric relaxation frequency. Ferroelectrics with much higher dielectric relaxation frequency would allow miniaturization and tunability of high-frequency microwave circuits.

New Materials For Tomorrow's Technologies While high-transition-temperature materials have attracted most of the scientific interest in superconductivity in recent years, most commercial superconducting devices use lower transition-temperature materials composed only of metals. Industry has invested heavily in oxide superconductors for passive microwave components, SQUID magnetic-field sensors, high-current conductors for power transmission and high-field magnets, and Josephson-based circuits for mixed-signal (analog and digital) circuits and AC voltage standards. The recently discovered borides are also receiving considerable attention.

While superconducting materials offer potential performance advantages over conventional materials, this has so far come at the cost of additional system complexity due to both the refrigeration requirement and the challenges posed by other properties of the materials, such as the brittleness and anisotropy of high-temperature superconductors. Unless refrigeration becomes efficient and reliable or higher temperature superconducting materials are discovered, superconducting devices will be relegated to niche military applications.

Speculative concepts such as quantum devices and spintronics have the potential to dramatically change the way military information is processed and encrypted. These concepts require exquisite control of material structure and properties on the nanometer if not subnanometer scale—a tremendous growth and processing challenge. The development of materials in the form of arrays of nanowires instead of thin films is a promising approach. New materials may also be needed. While the risks of such research are high, the possible payoffs for the first nation to field systems based on the successful embodiment of some of these concepts may be immeasurable. In the limit of miniaturization, it can be expected that electronic devices will reach the scale of small numbers of atoms or molecules—the domain of molecular electronics. The challenges associated with realizing this dream are discussed in Chapter 6.

The military will continue to have an insatiable appetite for data storage. The sorts of materials issues that arise in pushing storage density limit to the ultimate are essentially the same as for other nanostructured materials: the needs to control structure and composition on the atomic or nanometer scale, and to be able to read and write on that same scale. The specific materials required will also probably be different from today's.

Priorities for DoD Materials Research in Electronic Materials

The priorities for DoD-funded research into electronic materials fall into only a few categories:

Fundamental Understanding of Existing Materials Known materials, such as SiC and Group III-nitrides, are very promising for high-power and high-frequency applications. These materials are different from and in many respects more complex than the materials they would replace. Understanding the performance limits of the materials, as well as achieving them reproducibly, will require a multipronged scientific investigation incorporating experiment, theory, and modeling.

New Materials with Extreme Properties The need for a material with an electro-optic coefficient 500 times higher than traditional oxides to meet new requirements for ADCs is an example of extreme materials properties that will require new approaches to materials structure and processing. The desire for electronic materials that can function around 500°C is also an extreme demand. Room-temperature superconducting materials and quantum devices would profoundly change system options. Theory and modeling will be key in guiding the experimental approaches to promising compositions and structures.

Packaging and Thermal Management High-power devices that may function at temperatures around 500°C place severe requirements on interconnect, overlayer, and packaging materials and on the system components in close proximity to the device itself.

Materials Processing Optimizing the performance of existing materials and successfully fabricating new ones with extreme properties will succeed or fail in large measure with the emphasis given first to understanding and then to controlling their fabrication and processing. As an example, improved materials and processing will enable production of components and subsystems (Figure 5-2) for frequency-agile communication and radar

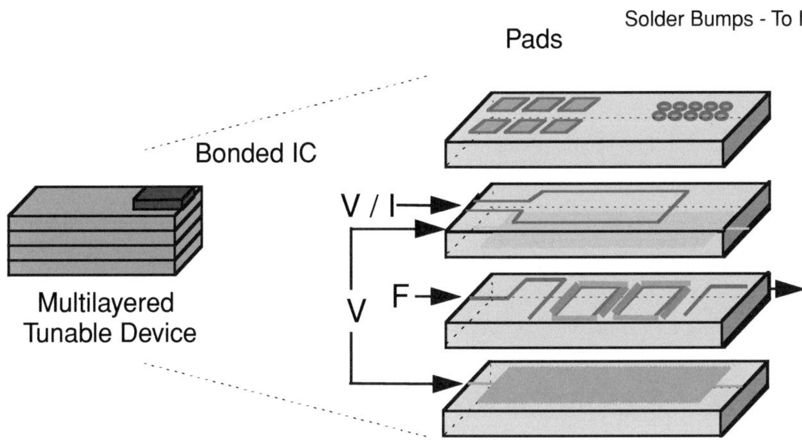

FIGURE 5-2

Generalized concept incorporating oscillators, filters, phase shifters, and circulators for a multilayer package with integrated circuits to improve quality and impedance matching. SOURCE: Courtesy of S. Wolf, DSO Office, DARPA.

applications that take advantage of the field-variable properties of ferroelectrics, ferrites, and other new materials.

Theory and Modeling Understanding the role of defects and the relative importance of defects in materials like SiC and GaN will depend heavily on theory. Elimination or control of the most detrimental defects is likely to be done experimentally, with significant guidance from growth models. In design of new materials with extreme properties, theory is likely to guide modelers to the most promising general structures and compositions. Modeling in turn will be required to narrow the list of most promising materials to a number that can be tested experimentally.

Materials for Revolutionary Technologies The scaling of both Si planar CMOS technology and high-density storage using known approaches will reach fundamental limits within the next 15 years. Perhaps the most fruitful area in which to look for potentially revolutionary technologies is in materials and material architectures that would enable entirely new approaches to computation, encryption, and data storage. A first step in this direction would be to find materials and processes that could be used to demonstrate experimentally some of the current theoretical concepts in quantum computing, spintronics, etc. (Gupta et al., 2001; Roukes, 2001). Materials concepts that may be relevant are:

- Integration of different functions on the nanometer or possibly atomic scale,
- Parallel material architectures that enable parallel function or fault tolerance,
- Self-assembly and self-repair, and
- Reconfigurability.

The integration of carbon nanotubes or biological molecules into material structures to perform some or all of these functions is an intriguing possibility.

Conclusions

The might and productivity of the commercial sector in silicon technology and in magnetic and optical storage technology dictate that the DoD leverage this technology base as much as possible. DoD investments should emphasize electronic materials for systems to identify, locate, and engage or defend against threats; for compact systems to transmit very high power at very high frequencies; and for embedded technologies that enable these systems. Materials research investments that are most likely to pay dividends are fundamental studies to increase understanding of existing materials; development of new materials with extreme properties; materials processing; packaging and thermal management; theory and modeling; and materials for revolutionary technologies.

Optoelectronics and Photonics

Commercial Drivers

The primary commercial drivers for photonic and optoelectronic materials are telecommunications, data communications, high-speed interconnects, optical data storage, light-emitting diode (LED) lighting, displays, and instrumentation. Telecommunications focus on InP and related materials operating at wavelengths around 1.55 mm. Datacom emphasizes high-volume and low-cost manufacturing using GaAs and InP materials. High-speed interconnects among electronic systems, boards, and chips, or even within chips, employ arrays of low-cost lasers and detectors using mainly GaAs-based materials, such as very short reach vertical cavity surface-emitting lasers. Heterogeneous integration on silicon substrates or organic substrates and even printed circuit boards is of interest to reduce cost and size. GaN and other wide-bandgap semiconductors are the main materials for optical data storage, LED lighting, and large panel display applications. There is no direct overlap between the commercial and DoD drivers in terms of materials.

DoD Drivers with Potentially High Materials Impact

The DoD drivers for future optoelectronic and photonic materials can be divided into three categories: (1) weapons, (2) electro-optic sensors, and (3) communications. The weapon applications include high-power lasers and infrared countermeasures (IRCM). Sensor applications include imaging at various bands from visible to very long infrared wavelengths, multispectral and hyperspectral imaging, terahertz imaging, laser radar, sensing of chemical and biological agents, and sensor/eye protection. In the communications area, DoD requires broadband coverage of RF signals with high fidelity. The material requirements for these applications are far beyond those of commercial drivers. The panel focuses its discussion on inorganic materials; organic photonic materials are discussed in Chapter 6.

High-Power Lasers Laser weapons require output power exceeding one kilowatt. Laser materials with high saturation power/fluence and high thermal and dielectric damage thresholds are key. Solid-state lasers are the main contenders for high-power lasers. For solid-state lasers, the primary challenge is the growth of large laser crystals with few defects or impurities.[3] Most current efforts focus on the growth of pure bulk crystals. Artificially engineered laser materials could increase the size and spatially vary properties. One example is fusion bonding of individual crystals to form larger crystals with dopant concentration gradients for uniform heat deposition and spatial mode improvement (Machida et al., 1998). Another is ceramic media created by hot-pressing powders of conventional laser materials to form large rods with tailorable concentration (Jianren et al., 2001).

[3]Hagan, D.J., "Overview of Photonic and Laser Materials," briefing presented to the Panel on Electronic and Photonic Materials of the Committee on Materials Research for Defense After Next, National Research Council, Irvine, CA, October 10, 2001.

Infrared Countermeasures Infrared seekers use different detectors to observe the IR signature of aircraft. The IRCM laser must cover every potential detector wavelength band with sufficient power. Optical parametric oscillators (OPOs) can efficiently shift laser wavelengths through the mid-IR region. Broad wavelength tunability is possible. The challenges are to develop new materials with both high nonlinearity and large power handling capability. Phase matching is critical for high conversion efficiency. Quasi-phase-matching (QPM) using periodically modulated materials is a powerful technique to force phase matching (Burr et al., 1997; Finsterbusch et al., 2000).

Clever engineering of composites of known materials may produce the best new materials. Several QPM techniques have been proposed, including stacking crystals of alternating orientations, periodic poling of polymers or fused silica, ferroelectrics, asymmetric quantum wells, and epitaxial growth of semiconductor materials on template substrates. Engineering of the grating could further tailor the properties of synthetic nonlinear materials. For example, multiple gratings can be used for broad tuning of OPOs, with each grating designed for a different wavelength. Aperiodic gratings can shape ultrafast femtosecond pulses similar to digital filters.

Another challenge of IRCM is replacing optical steering mirrors with electronic beam steering (as discussed briefly above). Electronic switching could be achieved by using dielectrics, piezoelectrics, photoconductors, or semiconductors. This will require design and production of film and bulk hybrid materials with properties well beyond those of traditional oxides. Magnetic antennas, antennas that can steer and tailor radiation patterns dynamically, and periodic dielectric structures that forbid propagation of electromagnetic waves in a certain range require novel bulk and thin-film crystals. MEMS micromirrors and electro-optic beam steering devices are also promising.

Materials with high damage threshold, fast modulation capabilities, and low absorption are required. Materials for phase shifters with extremely low loss at high frequency are essential. This will require the design and fabrication of tunable materials with low-loss dielectric and piezoelectric properties. Materials scientists have to look beyond the horizon of traditional materials to multinary materials. Possible approaches are fabrication of crystalline materials with a large number of different atoms and huge unit cells, or the combination of more traditional materials integrated in one, two, or three dimensions on the nanometer or micrometer scale. The synthesis and processing of such complex materials will be a daunting challenge.

Focal Plane Arrays and Multi- and Hyperspectral Imaging Hyperspectral imaging (HSI) sensors in the reflective region of the spectrum (sometimes referred to as imaging spectrometers) acquire digital images in many contiguous and very narrow (nominally about 0.010 μm wide) spectral bands in the visible, near-IR, and mid-IR portions of the spectrum (0.4–2.5 μm). This enables the construction of an essentially continuous radiance spectrum for every pixel in the scene. Thus, HSI makes possible the remote identification of ground materials of interest based on their spectra signatures. HSI has great potential for identifying camouflaged soldiers, trucks, tankers, and other military vehicles. It would be particularly attractive if systems could be included in army vehicles as well as in the equipment of an individual soldier.

HSI complements and enhances radar imaging. Major materials challenges are focal plane arrays covering ultraviolet (UV), IR, and very long wavelength IR regions. Current IR focal plane arrays are based on selenide- and telluride-based materials or inter-subband transitions in quantum wells (Dyer and Tidrow, 1998; Gunapala et al., 2000).

The semiconducing bandgap of carbon nanotubes (sub-100 meV to a few hundred meV) and the ability to control bandgap through tube diameter and chirality make them attractive for IR detection in the 1-15 μm range (Xu, 2001). Tuning is achieved by varying nanotube diameter. The cylindrical symmetry of the nanotubes allows absorption of all linearly polarized light, so that increasing the film thickness (i.e., tube length in a well-aligned film) increases absorption in a carbon nanotube-based detector. The major materials challenge here is to consistently control the parameters that govern nanotube electrical properties.

Electro-optic sensors for information-gathering and battlefield operations present similar challenges. There is a dearth of materials to meet the requirements of such systems. Materials with large acousto-optic merit will need to be designed, fabricated, and processed, requiring significant interplay among theory, modeling, experiment, and engineering. For small systems it is essential to have materials and the concomitant processes and packaging that can operate at UV, mid-, and far-IR wavelengths. Warfighter camera-type systems can achieve much higher resolution and range if better materials for light-gathering and detection are identified and developed.

Eye/Sensor Protection Optical limiters protect sensors from strong laser irradiation. Passive optical limiters can use optical processes such as absorption, refraction, or scattering. Materials with giant nonlinear optical properties are needed. Nonlinear photosensitive liquid crystals have been used in optical limiters (Khoo et al., 1998). Because the nonlinear optical coefficients scale inversely with the bandgap (to the third or fourth power), giant nonlinear effects in traditional bulk media are unlikely. Instead, nano-engineered materials are promising for increasing nonlinear effects. Nonlinear photonic crystals and nanostructured materials are also promising, as are metal-dielectric nanocomposites that can enhance the electrical field. Strong confinement of electrons in organic materials or quantum dots may lead to enhanced nonlinear optical coefficients. Porous sol-gel glasses with thermal index change could enhance nonlinear optical scattering. To protect the eyes of soldiers, the optical system needs to be lightweight and much less bulky. Nonlinear optical coatings would be ideal for this application. They would require an even larger (10^4!) enhancement of nonlinear optical properties and present tremendous materials challenges.

Terahertz Imaging The terahertz frequency range has many DoD applications (Mickan et al., 2000). The primary application is imaging, including satellite-to-ground. Terahertz radiation offers better resolution than millimeter waves and can detect covered targets (e.g., tanks under trees). It is also useful for detecting landmines, though it is limited to dry soil because water has high attenuation in the terahertz range. Terahertz radiation can image subsurface structures and is useful for airport security and medical diagnosis. Currently, the dominant techniques to generate terahertz radiation use optical rectification excited by femtosecond optical pulses. Optical rectification requires materials with large second-order nonlinearity. New materials such as nonlinear organic media or periodically poled lithium niobate and other artificially structured materials with enhanced efficiency are needed. On the receiver side, crystals with high nonlinearity, better phase matching between optical and terahertz waves, and low birefringence are needed.

Photomixing is another technique to generate terahertz waves whose frequency is equal to the difference frequency of two continuous-wave laser beams. High-efficiency and high-power photomixers are required. Materials such as low-temperature grown GaAs have been successful for short optical wavelength sources. New materials responding to long wavelengths (1.55 μm) could take advantage of rapid advances in commercial telecommunication technology. Ultimately, all-solid-state terahertz sources using inter-subband transitions in quantum wells, superlattices, or other bandgap-engineered semiconductors would be the most compact and energy-efficient sources (Williams and Hu, 2001). Most semiconductors, however, have phonon bands in these wavelength regions (8.8 THz for GaAs, and 10.4 THz for InP). New "phonon-engineered" materials may have to be created for this technique to become practical.

Chemical/Biological Sensing Photonics will continue to be a primary approach to sensing chemical and biological agents in the form of gases, vapors, aerosols, or surface-bound media. For chemical sensing there are several different techniques. The main material challenges are for new IR laser media. Compact semiconductor quantum cascade lasers (QCLs) are a recent breakthrough that extends the wavelength range of semiconductor lasers well into the mid-IR. The lasing wavelength of the QCL can be tailored by bandgap engineering, but operation is currently limited to cryogenic temperatures. For biosensing, UV fluorescence will likely remain a predominant detection technique because of its sensitivity and ubiquity among biological species. The lack of efficient UV laser and detector media are barriers to progress. Wide-bandgap semiconductors made from GaN are receiving the greatest attention in the engineering research community, but fundamental studies are lacking. These issues are discussed further in the section on microsystems.

Communications Broadband RF photonic links and systems are needed for transporting and processing signals from RF sensors and phased-array radars and to transmit signals to towed decoys. These systems require high dynamic ranges and large signal-to-noise ratios. Commercial developers of fiber optic networks are primarily concerned with digital signals. The material challenges for DoD are broadband electro-optic (EO) materials with high EO coefficient and high optical power-handling capability. Current materials like bulk EO crystals and semiconductors are approaching their limit. New materials based on photonic crystals could greatly enhance modulation efficiency. Though EO polymers are promising for broadband operations because of close index matching between optical and microwaves, their temperature range is very limited (Steier et al., 2001). New EO polymers that are stable at higher temperatures are desired.

Priorities for DoD Research in Optoelectronic and Photonic Materials

As the above discussion makes clear, most innovations in materials lie in clever engineering of materials in the nanometer (1 to 100 nm) range (nano-engineered materials). Nano-engineering refers to combining two or more different materials in one, two, or three dimensions. Nano-engineered materials encompass many different classes of materials, including photonic crystals, as shown in Figure 5-3 (dielectric/air or dielectric/metal at the 100-nm scale); electronic bandgap-engineered materials (one-dimensional semiconductor superlattices or heterostructures at 1- to 10-nm scales); quantum dots; nanowires; and many other novel materials. Combining these capabilities monolithically would enable realization of photonic and/or photonic/electronic device integration. Successful nano-engineering will depend critically on the guidance of theory and modeling to determine what structures should be made. Materials processing will then play a crucial role in fabricating reproducible structures with the required tolerances.

Photonic bandgap materials are promising for achieving significantly higher EO, nonlinear optical, and acousto-optic coefficients than bulk materials. These materials will affect many DoD applications, including IRCM, sensor/eye protection, and RF photonic communications. They promise to enhance the sensitivity of detectors over a specified wavelength range, which will have an impact on multispectral and hyperspectral imaging. They can also be used for IR camouflage of soldiers. Achieving the same level of control over photons in photonic bandgap materials that we currently take for granted for electrons and holes in semiconductors will require the teamwork of theorists, modelers, and experimentalists. Achieving photonic bandgap structures in all the wavelength regions of interest will be a grand challenge for materials processing.

Electronic bandgap engineering can tailor the electronic and optical properties of materials. It extends the wavelength coverage of lasers and detectors over that of conventional bulk or quantum well materials. QCLs extend the laser wavelength into the VLWIR region. IR detectors, focal plane arrays, and hyperspectral imaging could benefit from type II semiconductors or bandgap-engineered materials.

In addition to DoD encouraging nano-engineered materials, continued support for discovering new host materials is also important. These include semiconductor materials with larger bandgap (for UV lasers/detectors) and small bandgap (infrared focal plane array) and photonic materials with large EO, nonlinear optical, and acousto-optic coefficients for nonlinear and EO applications.

Heterogeneous materials at the microscale (1 to 100 μm) are important for photonic materials. Large laser crystals with tailorable dopant gradients can be formed by fusion bonding of individual crystals with different

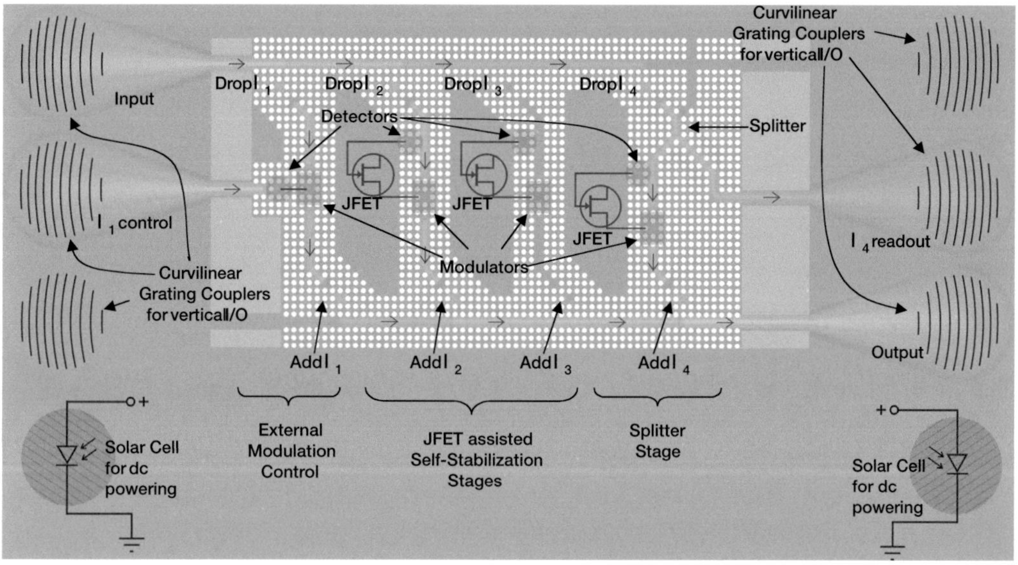

FIGURE 5-3

Schematic of nanoscopic photonic integrated circuits made of photonic crystals. Note that four WDM channels are shown, but that is meant to be representative of many more channels per chip. SOURCE: Courtesy of E. Yablonovitch, University of California, Los Angeles.

doping concentrations. Novel techniques to synthesize quasi-phase-matched materials will be useful for high-efficiency nonlinear materials.

Organic materials have improved tremendously in recent years in the variety of functions they can perform and in their aging and reliability characteristics. As discussed in Chapter 6, organic materials are likely to play a role (either alone or in concert with other materials) in achieving at least some of the functionality described here. The materials research needs for organic materials for optoelectronic and photonic applications is in many respects similar to the needs described here for inorganic materials.

In addition to the potentially revolutionary technologies discussed, such as photonic bandgap engineering, other concepts such as optical computing are very exciting but will depend to a large degree on advances in materials for their realization.

Conclusions

The major DoD drivers for future optoelectronic and photonic materials are for weapons, EO sensors, and communications. There is no direct overlap of materials needs between commercial and DoD drivers. While continual progress could be made by improving current materials, greater opportunities lie in nano-engineered materials with tailorable optical or EO properties. These include photonic bandgap materials, electronic bandgap-engineered materials, nanocomposites, and quasi-phase-matched materials. These materials are the most promising route to improving performance by orders of magnitude, as DoD applications require. In order for these materials to be exploited to the fullest, close collaboration between theorists, modelers, and experimentalists will be essential. Control of the structure and properties of materials on the nanoscale will present a tremendous challenge for materials processing.

Microsystems

Commercial Drivers

Microsystems and MEMS are subjects of continuing R&D; commercial products are established and new applications are emerging. Arrays of steerable micromirrors, for example, are used in computer projector systems and are available for optical communications and other applications. Accelerometers for triggering the deployment of automobile air bags are now based on MEMS technology, as are commercial ink-jet printers. A variety of simple "lab-on-a-chip" devices are commercially available, and new concepts for chemical analysis, drug discovery, medical diagnostic and other uses are being explored. Commercial development of environmental chemical sensing systems will be vigorous, second only to medical sensing in projected private efforts. Food-related sensing efforts are targeting mainly microbial agents that cause spoilage, with some attention to genetically modified organisms. Both will stimulate commercialization of sensing systems with features similar to those needed for military purposes: small systems targeted at a few specific species in a matrix of many other (dissimilar) species. However, specifications for analysis times, target analytes, and reliability that are unique to military applications will require a separate effort.

Commercial microsystems technology is at a relatively early stage; unlike in microelectronics, there are no established material or technology platforms. Current microsystems materials include silicon-based materials, polymers, and ceramics. Much of the technology now being exploited in the commercial sector was supported initially by DoD and other federal funding. Significant R&D efforts are also in Europe, Japan, and other foreign countries. Because the costs are very high, and the risks significant, it is unlikely that U.S. industry alone will be the major driver for new materials and technology. DoD has an opportunity to play a significant role here, both to meet its own needs and to enhance U.S. competitiveness in this important but infant technology.

DoD Drivers with Potentially High Materials Impact

Microsystems that comprise electronic or photonic devices integrated with structures and devices that expand functionality beyond data transmission and processing are beginning to show clear promise as a revolutionary technology. During the next 20 to 30 years, they will become increasingly important for a wide range of defense applications. Microsystems may incorporate MEMS and microoptoelectromechanical systems (MOEMSs). With advances in materials processing, nano-electromechanical systems (NEMSs) may be created that include new device behaviors that capitalize on their small size, reaching ultimately to the molecular scale.

Miniaturization and integration of complex functionality into microsystems will significantly improve performance compared to similar functions performed by existing macrosystems. Among the improvements, expected to be commensurate with the degree of miniaturization, are increased operating speed and ruggedness coupled with reduced power consumption and weight. Capabilities currently impossible in large-scale systems may be possible with microsystems, for example, ultrasensitive sensors or methods of displaying three-dimensional optical images by manipulating the complex reflectivity of surfaces with subwavelength resolution.

DoD will have an increasing need to rapidly identify and contradict a more diffuse and technologically varying class of threats to both civilians and combatants. Microsystems will be essential in counteracting such threats. Several drivers were considered in evaluating the DoD priorities for microsystems. The first is the creation with microsystems of distributed self-contained systems that sense, think, act, and communicate, to be widely distributed in fixed locations or readily carried by nontechnical civilian responders, soldiers, autonomous movable platforms, or piloted vehicles. A related driver is the need for rapidly collecting and communicating information on human activity and the environment. A third driver addresses the unique military defense needs for information security and identification. A fourth driver is the need to maintain the health and performance of defense personnel and equipment. These drivers have led us to consider the following applications where microsystems could contribute significantly to defense capabilities:

- Miniature mobile or distributed sensing capability;
- Rapid biometric identification of individuals;
- Security and encryption (e.g. locks, codes, and identification);
- Soldier health monitoring and remote medical diagnosis and treatment;
- Equipment monitoring, diagnosis, and failure prediction and analysis (see Chapter 3, the Structural Materials Panel report, for discussion of related issues);
- Dynamic camouflage, stealth, and altered electromagnetic signatures;
- High-density data storage and displays; and
- Miniaturized chemical processing: power generation, propulsion, and point-of-use materials fabrication.

Inherent in the microsystems approach is the demand that function be integrated into a self-contained unit. Such a system would include, for example, not only a sensor or detector but also a method for sampling from a solid, liquid, or gas as well as a power source and a method for processing, storing, and transmitting information. Advanced systems could be capable of directed motion and intelligent response to a threat, such as location and identification of a chemical source and initiation of countermeasures.

The small volume and broad functionality of integrated microsystems will require new combinations of material properties and new processing methods. Future military microsystems are likely to incorporate diverse materials, bringing to the fore issues of interfaces and materials compatibility. Such integration will pose similar challenges as methods of materials processing and fabrication of reduced dimension devices evolve. While silicon processing technology is mature enough for production of integrated electronic circuits, manufacturing complex integrated devices with additional functionality will require more diverse materials and processing methods. Military technologies are likely to advance in concert with advances in consumer products. For both applications, reduced cost and wider availablility of high-functionality systems are driving work on microsystems. Military applications will create new materials requirements, such as system lifetime or range of operating environments.

Microsystems will enable interaction with the environment in a variety of ways, coupled with information storage, processing, and communications. The panel envisions a class of devices that incorporates advanced electronic and photonic capabilities, such as high-speed data processing, laser/optical detection, and communication with other functions involving motion, material flow, chemical interaction, force transduction, and mass detection. General microsystem applications and enabling technologies include:

- High-sensitivity sensors (physical, chemical, and biological);
- Actuators for motion;
- Transduction and measurement of small forces and masses;
- Fluid control, including control of gas, liquid, and solid particulates;
- Optical systems for detection, modulation, beam steering, displays, and information processing;
- Data storage and processing;
- Custom or combinatorial materials synthesis; and
- Power generation.

The panel evaluated the potential for exploiting these capabilities for defense applications, with particular attention to areas of possible revolutionary enhancement of performance.

Materials engineered at the nanoscale may provide new properties to be exploited and used in microsystems, for example in ultra-high-sensitivity systems for detecting a broad spectrum of environmental parameters. Insight and inspiration can be obtained from emulating biological systems that have successfully attained the type of integration that the panel contemplates. Living creatures (some snakes, for example) have integrated highly sensitive sensors in an efficiently powered mobile unit that can not only detect IR signals but also analyze them and respond. Insects can detect trace compounds in the air or on a surface, follow the chemical trail to the source, and respond in appropriate ways, such as mating, laying eggs, or eating. Analogies to the functions we would like to emulate in artificial systems are evident.

Microsystems will rely on integration of power sources, such as harvesting ambient light or stored chemical or electrical power. Detectors and sensors will be integrated with electronic data storage and processing. Responses will be communicated to other units and human systems through electronic and photonic means.

Actuators for Motion MEMS, NEMS, and microsystems can move physical objects with dimensions from micrometers to the molecular scale. Moving components at the scale of hundreds of micrometers and below can be made by microfabrication processes based on lithography, etching, and deposition. These structures can be moved by electrostatic, thermal, or magnetic drive. The motion of these devices, possibly at high frequency or by the contact of material interfaces, raises issues of material wear, defect motion, and fatigue. In such small systems, mechanical properties and mechanical losses can be strongly influenced by surface effects or the inclusion and location of grain boundaries and defects. The fact that mechanical properties of materials may differ at the microscale must be considered in selecting materials for micro- and nanoscale systems. Molecules that generate motion, possibly extracted from or inspired by biological systems, are candidates for actuation of fine-scale motion.

Transduction and Measurement of Small Forces and Masses The converse effect to the actuation of motion and mass transport is the detection of motion, small forces, and mass. The forces on micro- and nanostructured materials can be detected by a variety of physical processes, such as the creation of a voltage on a piezoelectric material. Small changes in mass can be sensed by observing the resonant frequency shift of micro-oscillators. Modulated electron tunneling and deflection of a laser beam by a moving structure have also been used to detect small motions. Such approaches are used in commercial scanning probe microscopes, often using microfabricated cantilevers as the force or motion transduction elements. The optical, electrical, and piezoelectric—in addition to the mechanical—properties can be important for microstructures used in force transduction. Materials that can perform these functions must be integrated with other materials of different functionality.

Physical Sensors The transduction mechanisms mentioned above can be used to detect a variety of forces and physical properties of the environment. Miniaturized pressure sensors, based on measuring the deflection of a membrane, exist. Fluid flow can similarly be measured by detecting the forces imparted to a moving device in contact with a moving fluid. Magnetic fields can be detected in a variety of ways, including measuring the deflection of a mechanical cantilever incorporating a magnetic thin film. Accelerometers and gyroscopes can be realized in microsystems where the linear and angular acceleration can be measured as the force on a movable mass.

Chemical Sensors The large variety of possible target analytes and ambients (see Table 5-1) demands microsensor systems that can manipulate and modulate samples physically, chemically, and spatially and probe several aspects of their structure simultaneously or in rapid succession. Stand-alone vapor sensors, sensor arrays, or even instruments with single-parameter detection elements can screen for a few recognized threats but cannot meet ultimate performance demands. These call for systems in which pretreatment and separation functions play an integral role. Combining spectral detection with pretreatment is a common theme in most state-of-the art chem/bio microsystem development efforts (Frye-Mason et al., 2000; Rowe-Taitt et al., 2000; Xu et al., 2000; Eiceman et al., 2001; Hocker, 2001; Wise et al., 2001) and should remain central to future efforts. Future systems will require reductions in size, analysis time, detection limits, and power. Enhanced sample capture, pretreatment, analysis, identification, and quantification will require incorporation of innovative materials or combinations of materials into systems using micromachining and related MEMS, MOEMS, or NEMS technologies.

The design and selection of the sensor interface materials that interact with the target analytes are critically important. A rational approach to probing independent features of analyte structure by means of monolithic or integrated heteroarrays is of utmost importance. Fundamental interaction models are needed to guide the design of such multitransducer interface materials. Linear-solvation-energy-relationship (LSER) models (Grate, 2000; Hierlemann et al., 2001) provide an excellent example of a broadly applicable modeling paradigm. The tradeoffs between selectivity, reversibility, and stability of molecular recognition interactions (Hierlemann et al., 1999) must be recognized and incorporated into materials development strategies.

Polymers will figure prominently in future chemical sensing systems as sensor interface layers as well as for preconcentration adsorbents, separation membranes or phases, fluidic components, and perhaps even substrates. Rationally designed and synthesized polymeric sensor interfaces for

TABLE 5-1 Microsystems for (Bio)chemical Targets

Application	Description	Reference
Chemical warfare agent monitoring	Detection of nonbiological species; known nerve, blister, blood toxicants; environmental concentrations; personal-sphere concentrations; airborne as gas or vapor; aerosol-adsorbed; waterborne; surface-borne; sensing battlefield environments	SBCCOM[a]
Biological warfare agent monitoring	Detection of pathogenic organisms; viral, bacterial, fungal; environmental concentrations; personal-sphere concentrations; airborne as free or adsorbed aerosol; waterborne; surface-borne; battlefield environment sensing	Walt and Franz (2000), NIJ[b]
Explosives detection	Detection of landmines, water mines; direct detection of explosive chemicals; surrogate measures (package material, physical presence)	Dubey et al. (1999)
Combat identification	Chemical markers to discriminate friendly soldiers; enemies (previously tagged)	NRC (2001)
Trigger for protective measures	Alarm level for deploying/donning/doffing (automated) personal protective equipment; antidote administration	Army[c]
Dose monitoring	Measurement of biological fluids (internally or externally) to assess extent of internal exposure or physiologic effect of exogenous agents; trigger for protective measures or antidotes; early warning of adverse health effects	NRC (2001)
Enemy presence	Gas, vapor, aerosol composition signatures indicative of personnel or mechanized troop presence	SBCCOM[a]
Fire warning	Early warning of damage to vehicles, quarters, aircraft	Rose-Pehrsson et al. (2000)
Wound status	Early indicators of infection	NRC (2001)
Nuclear non-proliferation	Detection of byproducts of nuclear weapons production	CMC[d]

[a]SBCCOM. 2001. Soldier and Biological Chemical Command. Available online at <http://www2.sbccom.army.mil/hld/>. Accessed June 26, 2001.

[b]NIJ, National Institute of Justice. Available online at <http://www.ojp.usdoj.gov/nij/pubs.htm>. Accessed December, 2001.

[c]Army. 2001. U.S. Army Soldier Systems Center. Available online at <http://www.natick.army.mil/>. Accessed June 27, 2001.

[d]CMC. 2002. Cooperative Monitoring Center. Available online at <http://www.cmc.sandia.gov/>. Accessed January 29, 2002.

nerve agents (Grate et al., 1997; Levitsky et al., 2001) and explosives (Yang and Swager, 1998; Walt and Franz, 2000) have already yielded significant improvements in detection. These could serve as starting points for constrained combinatorial syntheses for further optimization.

Nanoscale device fabrication raises intriguing possibilities for new sensing modes. Nearly all current microsensing approaches rely on changes that occur within interfacial films over relatively long distances or that involve bulk phenomena. As interface domains shrink to the nanoscale and as measurements of interactions between discrete entities (e.g., single polymer strands and "a few" molecules) become analytically accessible, new sensing paradigms arise. Sensitivity and selectivity are likely to increase, although the magnitude remains to be seen. Control of temperature and other external variables will become even more important. This domain, now largely uncharted, may provide an opportunity for revolutionary advances.

Optical Sensors An alternative, complementary, detection strategy for chemical analysis is vibrational spectrometry. IR spectrometry in particular is capable of particle-density determinations, multicomponent vapor analysis, and structural identification. Emergent detection schemes include differential absorption LIDAR, frequency modulation spectroscopy, and heterodyne detection. Raman spectroscopy has similar capabilities. The inherent low sensitivity of IR absorbance by probed samples could be addressed by improvements in broadband/multiband IR sources, uncooled detectors, and optical pathlength enhancements or sample preconcentration (or other amplification schemes), as discussed below.

Microbolometer detectors, configured into focal-plane arrays (FPA), possibly as micromachined dense arrays (Tezcan et al., 1999), appear to be a viable near-term approach to IR sensing in the wavelength regions of interest (mainly 2-5 μm and 8-12 μm and secondarily 1.5-1.8 μm and 2.0-2.4 μm). Particularly useful would be thin-film CMOS-compatible materials with high-temperature coefficients of resistance that improve upon known materials and that can be patterned on suspended bolometer structures.

A broadband IR source with wavelength selection or a source providing several discrete critical wavelength emissions compatible with spatially resolved FPA detection would enable IR spectroscopy with low detection limits. Compact IR semiconductor lasers are needed—for example, the long-wavelength QCL. Increasing the operating temperature for higher power output would make miniaturized IR spectophotometry possible. Two-dimensional IR methods employing multiple-frequency modulations and heterodyne detection are just emerging. With proper temperature control, they could permit unprecedented analysis of structural features. Path length or other concentration enhancements will be necessary for a microspectrometer to be effective for detection of trace (i.e., sub-ppm) analytes by IR spectroscopy. As an example, surface-enhanced Raman spectroscopy (SERS) realizes increases in sensitivity of several orders of magnitude (Stokes et al., 1999; Sylvia et al., 2000).

Biological Sensors DoD needs sensors for biological agents and pollutants. State-of-the-art bioaerosol sensing arrays, based on immunochemical or polymerase chain reaction (PCR)-amplified DNA-fragment fluorometry, have similar limitations as well as false positives, reliance on reagents, and time-consuming sample-handling and assay procedures (Walt and Franz, 2000). Bacteria, viruses, and many bacterial toxins, as well as all known nutrient media for biological agents, have strong UV fluorescence spectra that can serve as signatures for remote biodetection. Efficient UV laser and detector media are largely materials limited. GaN-based wide-bandgap semiconductors are promising for blue and UV lasers, although a fundamental understanding of the material is still lacking. The lack of a lattice-matched substrate presents a challenge for epitaxial growth of high-quality crystals.

Specific biological sensing operates as a lock-and-key system where a set of intermolecular interactions is defined by the geometry of a receptor site and a complex target molecule. Recent studies of cell-surface interactions have shown that short peptide sequences can be specifically recognized. Likewise, supramolecular chemistry is now showing that several hydrogen bonds between two molecules of constrained geometry are sufficient for a strong and specific bonding interaction. Such interactions are readily accessible in solution-phase systems but less so in gas-solid systems.

Many synthetic sensing systems are based on immobilized antigens or antibodies where binding is detected because of a mass change or other physical effect. Many sensors are also based on immobilized enzymes, which can be used to determine the concentration of substrate through pH change or redox reaction. A number of research directions are appropriate for developing specific sensors. Molecular modeling of intermolecular interactions is needed to enable the design of specific binding systems, to make a lock for any given key, and to define the necessary properties for a specifically detectable key.

Biosensors will require materials and assembly processes for thermal and fluid support systems for living cells. In addition, these sensors will need to be processed monolithically with the rest of the microsystem (Table 5-1). The biological aspects of these challenges are discussed in the report of the National Research Council Committee on Opportunities in Biotechnology for Future Army Applications (NRC, 2001).

Sensor Response Amplification Through Material Property Transitions Amplification forms a complementary theme for interface material development, as discussed in the Bioinspired and Bioderived Materials report (Chapter 7). While there are many biological examples, nonbiological amplification mechanisms also deserve attention. The realignment and reorientation of liquid-crystalline (LC) domains due to the interaction of an analyte with the anchoring site of an immobilized LC is a good example of this phenomenon (Shah and Abbott, 2001). Large changes in optical (and probably other) properties occur due to the long-range reorientation in a material caused by a single interaction site disruption. Another example is the use of photonic-lattice diffraction gratings with vapochromic materials (Bailey and Hupp, 2002), where refractive index changes are selectively enhanced by orders of magnitude near the visible-light absorption maxima of the interface material. Changes in polymer modulus contribute significantly to surface-acoustic-wave (SAW) sensor responses, calling for design of polymers that capitalize on this feature to enhance sensitivity (Grate and Zellers, 2000). Similarly, thickness-shear-mode (TSM) and SAW sensors can reach resonance conditions where responses become discontinuous, a possible source of an order-of-magnitude increase in response. Operating near phase transitions or conduction percolation thresholds represents other sources of response amplification. Nano-engineered materials with so-called hard (rigid) and soft (flexible) regions are an attractive option for designing in such properties. Guidance from theory and modeling will be essential.

Sample Manipulation, Fluid Control, and Thermal Management Microsystems for chemical and biological sensors or reactors (see below) require the transport and manipulation of fluids. Preconcentration, chemical reaction, and separation will generally require active mass transport. In liquid-phase systems employing aqueous media, this can be achieved with voltage biases at various points in a system. Gas systems relying on suction- or pressure-driven mass transport using microscale diaphragm pumps are being considered for microsystem applications but scale unfavorably with miniaturization.[4] The flexural-plate-wave approach (Moroney et al., 1991; Meng et al., 2000), whereby gas flow could be achieved by acoustic streaming, has been suggested but not fully explored. Distributed transport mechanisms based on ciliary, peristaltic, or other biomimetic actuation could be options for gas transport. Efficient low-power valves and pumps for both gas and liquids require chemically inert thin-film materials.

Preconcentration and focusing of samples, perhaps selectively, can be achieved with porous materials of adjustable surface area, pore size, and functionality. Common but proprietary materials used for this purpose are carbon molecular sieves, graphitized carbons, and rigid polymers, which are thermally stable to over 400°C. Adapting these or similar types of materials, to allow in situ deposition, patterning, and control of porosity within etched channels of a microsystem would be useful for gas-phase and liquid-phase analysis. Tailored organic/inorganic nanocomposites are an attractive class of nanoengineered materials for this application. Templating by incorporation of porogens of selected sizes or functionality can impart a crude level of selectivity in these supramolecular systems (Raman et al., 1996).

Molecular separation based on controlled motion, competitive binding, or affinity would also benefit from nanoengineered materials and micromachined channels. Imparting structural features that would allow modulation of the binding or affinity with electric or photonic input would be a novel means of providing in situ control of separations. Clever channel construction methods and architectures that build upon anisotropically etched Si (Matzke et al., 1998), vapor-phase deposited polymers,[5] and sacrificial-polymer processing (Mastrangelo et al., 1998; Bhusari et al., 2001) are examples of current research. Hybrid organic-inorganic nanocomposites are also promising (Lu et al., 2001).

---

[4]Cabuz, C. 2001. Available online at <http://www.darpa.mil/dso/thrust/md/Mm/pump/honeywell/html>. Accessed March 29, 2001.

[5]Noh, H.-S., C. Bonner, P.J. Hesketh, and G.C. Frye-Mason, "Fabrication of Parylene Column for Micro Gas Chromatograph," paper presented at the International Symposium on Mechatronics, Atlanta, GA, 2000.

Finally, heating and cooling elements are needed to control the temperature of certain systems components. Microfluidic systems may provide thermal management. Alternatively, thermoelectric (TE) materials offer a promising approach. Unfortunately, thermal desorption and temperature-programmed separations require heating and cooling rates and values that exceed the capabilities of current TE materials. Materials or approaches that reduce demands for high temperature or that allow high temperatures to be achieved at low power are required. Fluid motion could also be used for control of the optical or mechanical properties of devices.

Other Novel Applications Important R&D activity is under way on MEMS devices and microsystems for optical devices, data processing, data storage, and displays, motivated by both defense and consumer applications such as telecommunications. MEMS-based systems for chemical and biological sensing are also being pursued vigorously.

With the opportunities to develop new sensor and actuator microsystems come significant materials challenges. The same types of fluid control systems being developed for sensors may also be exploited in microsystems for controlled chemical reactions. These chemical reactor systems could be used to generate power, for example, through internal combustion or fuel-cell-type reactions. These power sources could integrate with microsystems or be used for other compact power applications, as discussed in Chapter 4. Miniaturized or highly parallel chemical reaction systems could also be used for combinatorial approaches to fabricating novel materials for study or for creating materials that might be difficult or impossible to fabricate in bulk. Highly miniaturized reaction systems even approaching the nanoscale may provide a path to effective nanomaterials fabrication and processing, a need identified in Chapter 6.

Priorities for DoD Materials Research on Microsystems

The panel envisions microsystems with more diverse materials than current electronic, photonic, sensor, or structural systems. Materials for signal processing, power transmission, and communication will need to be integrated with materials that provide structural, chemical, or mechanical functions. Compatibilities among diverse materials, particularly at interfaces, will be increasingly challenging. As component dimensions shrink and nanoscale materials are employed, surface/volume and interface/volume ratios will be large. Methods to interface nanoscale (grown) materials with engineered devices will grow in importance. Issues like thermal expansion mismatch and interdiffusion must be mitigated.

Exploitation of nanoscale phenomena and even connection of biological or bioderived materials may be a valuable direction, because biological systems have arrived at solutions to many of the functionality issues we hope to address. The effective incorporation of bioderived or bioinspired materials into durable engineered systems remains a challenge. Nanoscale elements, perhaps carbon nanotubes, or functional macromolecules with nanoengineered properties may provide new material options. This raises the need to integrate molecular components into functional systems.

One of the fundamental areas for significant research will be on compatible materials for the control, metering, and flow of fluids. The panel envisions mass flow control with capabilities resembling those of electrical charge and light-flow control in electronic and photonic systems. Similarly important will be the ability to actuate and detect the motion of mechanical components in fluid and dirty environments. Chemical resistance and compatibility are likely to be critical, and there is strong motivation to incorporate into integrated systems materials with broad chemical resistance. Thermal management and temperature stability will also be important.

Fabrication of Microsystems In fabricating microsystems, one can build on the processes already well developed for microelectronic and photonic device and substrate fabrication. Lithographic processes created for the semiconductor industry can be exploited with the addition of etching and growth processes specific to the broader range of microsystem materials. Hybrid arrays of discrete sensors have been reported (Haug et al., 1993) and even commercialized (Gardener and Bartlett, 1999), but have not yet been fully exploited. With the possible exception of acoustic-wave sensors requiring a piezoelectric transduction layer, current sensor substrate materials are not performance-limiting. Furthermore, they are compatible with CMOS processing (Hagleitner et al., 2001), permitting integration of heteroarray sensors with each other and with Si-based circuitry and MEMS components. Such a monolithically integrated heteroarray could provide a wide variety of sensors with electronics and a full range of MEMS capability, but fabricating such a complex microsystem presents tremendous materials processing challenges.

The capability of controlling the location of polymers with different properties within microscopic tolerances throughout a microsystem needs to be developed for microanalytical systems. In situ thermal or photopolymerization from liquid- or, preferably, gas-phase precursors is one promising approach (Healey et al., 1995; Prucker et al., 1999; Hsieh and Zellers, 2002; Yu et al., 2001). Nanometer-resolution soft lithography (Huck et al., 2001), laser-induced manipulation and deposition of nanoscale particles (Niidome et al., 2001), and matrix-assisted laser desorption (Pique et al., 2000) are other novel approaches. Further technique developments and the design of materials for deposition, patterning, in situ modification, and adhesion could yield significant payoffs.

Microsystem Packaging The packaging of electronics continues to be an area demanding new materials. For microelectronics, the drive to higher densities, higher speeds, and smaller sizes increases the need for thermal management. Military devices and sensor-based microsystems may be required to operate in hostile environments, placing even greater demands on electronics packaging. Materials are needed that are mechanically durable, thermally stable, and impervious to gas or chemicals that could alter the devices. The packages must also allow for low-loss electrical connections. Photonic devices must connect to guided-wave or free-space optics, requiring materials with transparency across various parts of the electromagnetic spectrum. Sensor-based microsystems create additional demands on materials; they must be nonmagnetic, chemically inert, and even biocompatible. These microsystems also require interconnections for mass transport and storage of biochemical materials. Low-temperature processing and joining methods must be developed that are compatible with the new materials. The demand for new materials and processes for electronic, photonic, and microsystem packaging is very strong; lack of appropriate materials may limit the functionality of future systems.

DoD will need packaging methods that allow for systems to interact with the environment in ways—e.g., exchanging fluids, chemical communication, and optical coupling—that protect components that are environmentally sensitive. Packaging must also be robust, enabling device operation in environments as hostile as inside a human body, an operating engine, or a high-speed vehicle.

Conclusions

Microsystems are already showing their promise as a revolutionary technology that gives future U.S. military a significant advantage. The advantages of creating low-power, compact, lightweight, reliable systems for defense applications are compelling. They motivate addressing the significant materials issues involved. Although substantial R&D activity is under way, attention is particularly needed to materials processing, incorporation of bioderived and bioinspired components, materials compatibility, integration of nanoscale materials with engineered devices, and packaging. In the near term, emphasis is likely to be on integrating known materials into microsystems in a way that preserves the functionality of each material. In the future, new materials will be needed to increase system functionality. Power and thermal management are critical cross-cutting issues, with ultimate links to the materials used in electronic and photonic engineered systems. Mass flow control and chemical compatibility are emerging needs for many microsystem applications.

Rationally designed polymeric, nanocomposite, and anisotropic materials assembled or implemented in a way that allows response amplification, show significant promise as chemical microsensor interface materials and as preconcentration and separation media. Models to guide the design of such materials for multisensor heteroarrays are needed to optimize their information yield. Nanoscale devices that rely on quantum effects represent the most intriguing and potentially revolutionary avenues for research, with likely improvements in detection sensitivity and some promise for improved selectivity. Coupling materials design, synthesis, and film deposition strategies with MEMS and MOEMS device and system fabrication strategies is essential to mass production and to ensuring compatibilities during bonding, interconnecting, and packaging.

RESEARCH AND DEVELOPMENT PRIORITIES

This panel considered a wide variety of military needs from several vantage points, ranging from individual devices or components to entire (miniature) systems. Six common themes emerged that point to the most important areas of research.

Fundamental Understanding of Existing Materials

Important properties of many materials that are promising for defense applications are not well understood. While empirical engineering of known materials for a specific application can be expected to gradually advance the state of these materials, this approach has limits. The most dangerous possibility, perhaps, is that an undiscovered intrinsic property of a material may ultimately make it wholly unsuitable for the application. On a more optimistic note, fundamental scientific understanding is likely to give knowledge of the ultimate properties of a material and serve as a guide to its optimal processing. Truly fundamental understanding of complex real-world materials is only possible through a judicious combination of experiment, theory, and modeling, with intense feedback among the three. This will become even more important as we explore materials on the nanoscale at the interface between the discrete (quantum) and continuum (bulk) regimes of material behavior, where widely applicable modeling tools have not yet been developed.

New Materials with Extreme Properties

The increased functionality integrated into ever-decreasing volume drives the search for new materials that have properties (e.g., nonlinear optical properties) that are enhanced by an order of magnitude or more over known materials. Comparable improvements in material purity are also likely to be needed. Nanoengineered materials whose structure, composition, and morphology are varied on a nanometer scale in one, two, or all three dimensions are probably required.

Another approach that is appropriate in some cases is to search for single-phase materials with more elements and crystallographic complexity than have historically been considered. Identifying such promising materials is daunting. Effective and efficient experimentation will require theoretical and especially computational tools that can deal with such complex systems reliably. Optimization and exquisite control of processing will be essential to the reproducibility of such complex structures, especially those that combine different classes of materials (e.g., organic and inorganic).

New Ways of Combining Materials

Another way of increasing functionality in a small volume is to combine materials in new ways. Routine incorporation of nanoengineered components into a system (especially a microsystem) would both add functionality and reduce volume and weight.

Different classes of materials need to be combined in ways that are only beginning to be studied. These combinations often use inorganic materials that have high processing temperatures, structural rigidity, and known long-term stability with organic or even biological materials that have low temperature tolerances, relatively poor structural stability, sensitivity to the ambient (which is usually why they were chosen in the first place), and poor or unknown long-term stability. Processing such hybrid structures presents a tremendous challenge. The interfaces between materials must be understood, because the functionality often is derived from materials properties very close to the interface. As device and system dimensions shrink, interfaces become even more dominant because an increasingly higher percentage of atoms or molecules find themselves at one interface or another. Theory and modeling can be expected to serve as guides to selecting appropriate combinations of materials and processing methods.

Packaging and Thermal Management

Electronic and photonic materials will be integrated at increasingly greater scales in the coming decades. While the functionality that such systems provide will be staggering by today's standards, it comes at a price. Packaging and thermal management will be a significant challenge. Research in these areas needs to be done hand in hand with research in the areas already outlined.

Materials Processing

Before they can be introduced into new systems, it will be necessary to understand, optimize, and control the processing of the materials discussed in this chapter. Fabricating such complex materials in the first place will be a formidable challenge, as will be the design of processes to incorporate these new materials into structures and devices for military systems. Low-temperature processes are desirable in most cases, indispensable in some.

The complexity contemplated is too great to be able to rely on empirical process optimization alone. Processing models built on a solid theoretical underpinning will be key. These models will need to span length scales from nanometers to the scale of the device.

Theory and Modeling

Every materials research need discussed in this chapter points to the critical role that theory and modeling will play. These tools are not, however, currently sufficiently advanced to provide what is needed. The linking of theory, modeling, and experiment in each paragraph above illustrates the need for all three to be advanced in concert over the coming decade. One essential is bridging the length scales, ranging from the nanoscale on which many of these materials will be structured to at least the device scale. It will also be critical to span time scales, from the times at which the materials respond (as short as picoseconds) to the length of time the material is expected to be deployed (years). This is truly a grand challenge to materials science.

REFERENCES

Bailey, R.C., and J.T. Hupp. 2002. Large-scale resonance amplification of optical sensing of volatile compounds with chemoresponsive visible-region diffraction gratings. J. Am. Chem. Soc. 124(23):6767-6774.

Bhusari, D., H.A. Reed, M. Wedlake, A. Padovani, A.A. Bidstrup-Allen, and P.A. Kohl. 2001. Fabrication of air-channel structures for microfluidic, microelectromechanical, and microelectronic applications. Journal of Microelectromechanical Systems 10(3):400-408.

Burr, K.C., C.L. Tang, M.A. Arbore, and M.M. Fejer. 1997. Broadly tunable mid-infrared femtosecond optical parametric oscillator using all-solid-state-pumped periodically poled lithium niobate. Optics Letters 22(19):1458-1460.

CMC (Cooperative Monitoring Center), Sandia National Laboratories. Undated. Available online at <http://www.cmc.sandia.gov/>. Accessed December 11, 2001.

Dubey, A.C., J.F. Harvey, J.T. Broach, and R.E. Dugan, eds. 1999. Detection and Remediation of Mines and Minelike Targets IV. Bellingham, WA: SPIE.

Dyer, W.R., and M.Z. Tidrow. 1998. Applications of MCT and QWIP to ballistic missile defense. Proceedings of the SPIE—The International Society for Optical Engineering (Infrared Detectors and Focal Plane Arrays V, Orlando, FL, 14-17 April 1998). SPIE-Int. Soc. Opt. Eng. 3379:434-440.

Eiceman, G.A., E.G. Nazarov, J.E. Rodriguez, and J.A. Stone. 2001. Analysis of a drift tube at ambient pressure: Models and precise measurements in ion mobility spectrometry. Rev. Sci. Instrum. 72:3610-3621.

Finsterbusch, K., R. Urschel, and H. Zacharias. 2000. Fourier-transform-limited, high-power picosecond optical parametric oscillator based on periodically poled lithium niobate. Applied Physics B (Lasers and Optics) B70(6):741-746.

Frye-Mason, G., R. Kottenstette, P. Lewis, E. Heller, R. Manginell, D. Adkins, G. Dulleck, D. Martinez, D. Sasaki, C. Mowry, C. Matzke, and L. Anderson. 2000. Hand-held miniature chemical analysis system (μChemLab) for detection of trace concentrations of gas-phase analytes. Pp. 229-232 in Proceedings on Micro Total Analysis Systems 2000, A. van den Berg, W. Olthuis, and P. Bergveld, eds. Dordrecht, The Netherlands: Kluwer.

Gardener, W., and P.N. Bartlett. 1999. Electronic Noises: Principles and Applications. Oxford, UK: Oxford University Press.

Grate, J.W. 2000. Acoustic wave microsensor arrays for vapor sensing. Chemical Reviews 100:2627-2648.

Grate, J.W., and E.T. Zellers. 2000. The fractional free volume of the sorbed vapor in modeling the viscoelastic contribution to polymer-coated surface acoustic wave vapor sensor responses. Anal. Chem 72:2861-2868.

Grate, J.W., S.N. Kaganove, S.J. Patrash, R. Craig, and M. Bliss. 1997. Hybrid organic/inorganic copolymers with strongly hydrogen-bond acidic properties for acoustic wave and optical sensors. Chemistry of Materials 9:1201-1207.

Guilbert, L., J.P. Salvestrini, P. Kolata, F.X. Abrial, M.D. Fontana, and Z. Czapla. 1998. Optical characteristics of triclinic rubidium hydrogen selenate. Journal of the Optical Society of America B—Optical Physics 15(3):1009-1016.

Gunapala, S.D., S.V. Bandara, A. Singh, J.K. Liu, B. Rafol, E.M. Luong, J.M. Mumolo, N.Q. Tran, D.Z.-Y. Ting, J.D. Vincent, C.A. Shott, J. Long, and P.D. LeVan. 2000. 640x486 long-wavelength two-color GaAs/AlGaAs quantum well infrared photodetector (QWIP) focal plane array camera. IEEE Transactions on Electron Devices 47(5):963-971.

Gupta, J.A., R. Knobel, N. Samarth, and D.D. Awschalom. 2001. Ultrafast manipulation of electron spin coherence. Science 292(5526):2458.

Hagleitner, C., A. Hierlemann, C. Lange, A. Kummer, N. Kerness, O. Brand, and H. Baltes. 2001. Smart single-chip gas sensor microsystem. Nature 414(6861):293-296.

Haug, K., D. Schierbaum, G. Gauglitz, and W. Gopel. 1993. Chemical sensors based upon polysiloxanes: Comparison between optical, quartz microbalance, calorimetric, and capacitance sensors. Sensors and Actuators B 11:383-391.

Healey, S., E. Foran, and D.R. Walt. 1995. Photodeposition of micrometer-scale polymer patterns on optical imaging fibers. Science 269:1078-1080.

Hierlemann, A., A.J. Ricco, K. Bodenhoffer, and W. Gopel. 1999. Effective use of molecular recognition in gas sensing: Results from acoustic wave and in situ FTIR measurements. Anal. Chem. 71:3022-3035.

Hierlemann, A., E.T. Zellers, and A.J. Ricco. 2001. Use of linear solvation energy relationships to model acoustic wave vapor sensor responses. Anal. Chem. 73:3458-3466.

Hocker, B. 2001. The polychrometer: A MEMS-based optical correlation spectrometer. Available online at <www.darpa.mil/mto/mems/summaries/Projects/individual_32.html>. Accessed November 10, 2001.

Hsieh, M.D., and E.T. Zellers. 2002. In situ UV-photopolymerization of gas phase monomers for microanalytical system applications. Sensors and Actuators B 82(2-3):287-296.

Huck, W.T.S., A.D. Stroock, and G.M. Whitesides. 2001. Synthesis of geometrically well defined, molecularly thin polymer films. Angew. Chem. Int. Ed. Engl. 39(6):1058-1065.

Jianren L., T. Murai, K. Takaichi, T. Uematsu, K. Misawa, M. Prabhu, J. Xu, K. Ueda, H. Yagi, T. Yanagitani, A.A. Kaminskii, and A. Kudryashov. 2001. 72 W Nd:$Y_3Al_5O_{12}$ ceramic laser. Applied Physics Letters 78(23):3586-3588.

Khoo, L.C., B.D. Guenther, Min-Yi Shih, P.H. Chen, and M.V. Wood. 1998. Optical limiting of submicrowatt long-pulse cw lasers with extremely nonlinear nematic liquid crystal films. Pp. 374-375 in Technical Digest: Conference on Lasers and Electro-Optics, San Francisco, CA. Washington, DC: Optical Society of America.

Levitsky, I., S.G. Krivoshlykov, and J.W. Grate. 2001. Rational design of a nile red/polymer composite film for fluorescence sensing of organophosphonate vapors using hydrogen-bond acidic polymers. Anal. Chem. 73:3441-3448.

Lu, Y., Y. Yang, A. Sellinger, M. Lu, J. Huang, H. Fan, R. Haddad, G. Lopez, A.R. Burns, D.Y. Sasaki, J. Shelnutt, and C.J. Brinker. 2001. Self-assembly of mesoscopically ordered chromatic polydiacetylene/silica nanocomposites. Nature 410(6831):913.

Machida, H., N. Iehisa, and N. Karube. 1998. 2.5 kW slab Nd:YAG laser using a 400 mm-long crystal bonded by optical contact method. Proceedings of the SPIE—The International Society for Optical Engineering (Laser Processing of Materials and Industrial Applications II, Beijing, China, September 16-19, 1998). SPIE-Int. Soc. Opt. Eng. 3550:469-476.

Mastrangelo, C.M., M.A. Burns, and D.T. Burke. 1998. Microfabricated devices for genetic diagnostics. Proc. IEEE 86:1769-1787.

Matzke, C.M., R. Kottenstette, S.A. Casalnuovo, G.C. Frye-Mason, M.L. Hudson, D.Y. Sasaki, R.P. Manginell, and C.C. Wong. 1998. Microfabricated silicon gas chromatographic micro-channels: Fabrication and performance. Pp. 262-268 in Proceedings of the 1998 SPIE Conference on Micromachining and Microfabrication Process Technology IV, September, Santa Clara, CA. Bellingham, WA: SPIE

Meng, A.H., N.-T. Nguyen, and R.M. White. 2000. Focused flow micropump using ultrasonic flexural plate waves. Biomedical Microdevices 2(3):169-174.

Mickan, S., D. Abbott, J. Munch, X.-C. Zhang, and T. Van Doorn. 2000. Analysis of system trade-offs for terahertz imaging. Microelectronics Journal 31(7):503-514.

Moroney, R.M., R.M. White, and R.T. Howe. 1991. Microtransport induced by ultrasonic lamb waves. Appl. Phys. Lett. 59:774-776.

Niidome, Y., A. Hori, H. Takahashi, Y. Goto, and S. Yamada. 2001. Laser-induced deposition of gold nanoparticles onto glass substrates in cyclohexane. Nano Letters 1:365-369.

NIJ (National Institute of Justice). National Institute of Justice Publications. Undated. Available online at <http://www.ojp.usdoj.gov/nij/pubs.htm>. Accessed December 12, 2001.

NRC (National Research Council). 2001. Opportunities in Biotechnology for Future Army Applications. Washington, DC: National Academy Press.

Pique, A., D.B. Chrisey, J.M. Fitzgerald, R.A. McGill, R.C.Y. Auyeung, H.D. Wu, V. Nguyen, S. Lakeou, R. Chung, and M. Duignan. 2000. Direct writing of electronic and sensor materials using a laser transfer technique. J. Mater. Res. 15:1872-1875.

Prucker, O., C.A. Naumann, J. Ruhe, W. Knoll, and C.W. Frank. 1999. Photochemical attachment of polymer films to solid surfaces via monolayers of benzophenone derivatives. J. Am. Chem. Soc. 121:8766-8770.

Raman, N.K., M.T. Anderson, and C.J. Brinker. 1996. Template-based approaches to the preparation of nanoporous silicas. Chem. Mater. 8:1682-1701.

Rose-Pehrsson, S.L., R.E. Shaffer, S.J. Hart, F.W. Frederick, D.T. Gottuk, B.D. Strhlen, and S.A. Hill. 2000. Multi-criteria fire detection systems using a probabilistic neural network. Sensor Actuator B 69(3):325-335.

Roukes, M. 2001. Electronic in a spin. Nature 411(6839):747.

Rowe-Taitt, C.A., J.W. Hazzard, K.E. Hoffman, J.J. Cras, J.P. Golden, and F.S. Ligler. 2000. Simultaneous detection of six biohazardous agents using a planar waveguide array biosensor. Biosensors and Bioelectronics 15:579-589.

SBCCOM (Soldier and Biological Chemical Command). Undated. Available online at <http://www.sbccom.apgea.army.mil/>. Accessed September 16, 2002; Homeland Defense. March 6, 2002. Available online at <http://hld.sbccom.army.mil/>. Accessed September 16, 2002; Defense Threat Reduction Agency. Undated. Available at <http://www.dtra.mil/>. Accessed September 16, 2002.

Semiconductor Industry Association (SIA). 2001. The International Technology Roadmap for Semiconductors, San Jose, CA. Available online at <http://public.itrs.net/Files/2001ITRS/Home.htm>. Accessed March 3, 2002.

Shah, R.R., and N.L. Abbott. 2001. Principles or measurement of chemical exposure based on recognition-driven anchoring transitions in liquid crystals. Science 293(5533):1296-1299.

Steier, W.H., M.-C. Oh, H. Zhang, A. Szep, L.R. Dalton, C. Zhang, H.R. Fetterman, D.H. Chang, H. Erlig, and B. Tsap. 2001. Recent advances in low voltage, high frequency polymer electro-optic modulators. Pp. 1-3 in OFC 2001: Optical Fiber Communication Conference, Anaheim, CA.

Stokes, D.L., A. Pal, V.A. Narayanan, and T. Vo-Dinh. 1999. Evaluation of a chemical vapor dosimeter using polymer-coated SERS substrates. Analyt. Chim. Acta 399(3):265-274.

Sylvia, J.M., J.A. Janni, J.D. Klein, and K.M. Spencer. 2000. Surface enhanced Raman detection of 2,4-dinitrotoluene impurity vapor as a marker to locate landmines. Anal. Chem. 72:5834-5840.

Tezcan, D.S., F. Kocer, and T. Akin. 1999. An uncooled microbolometer infrared detector in any standard CMOS technology. Pp. 610-613 in International Conference on Solid-State Sensors and Actuators (Transducers '99), Sendai, Japan, M. Esashi, ed. Tokyo: IEEE.

Walden, R.H. 1999. Analog-to-digital converter survey and analysis. IEEE Journal on Selected Areas in Communications 17:539-550.

Walt, D.R., and D.R. Franz. 2000. Biological warfare detection. Anal. Chem. Vol. 739A-746A.

Williams, B.S., and Qing Hu. 2001. Optimized energy separation for phonon scattering in three-level terahertz intersubband lasers. Journal of Applied Physics 90(11):5504-5511.

Wise, K.D., K. Najafi, D.M. Aslam, R.B. Brown, J.M. Giachino, L.C. McAfee, C.T.-C. Nguyen, R.O. Warrington, and E.T. Zellers. 2001. Wireless integrated microsystems (WIMS): The coming revolution in the remote gathering of information. Technical Digest for SensorsExpo Conference, Chicago, IL.

Xu, J., W.B. Whitten, and J.M. Ramsey. 2000. Space charge effects on resolution in a miniature ion mobility spectrometer. Anal. Chem. 72:5787-5791.

Xu, J.M. 2001. Highly ordered carbon nanotubes arrays and IR detection. Infrared Physics and Technology 42:485-491.

Yang, J.-S., and T.M. Swager. 1998. Fluorescent porous polymer films as TNT chemosensors: Electronic and structural effects. J. Am. Chem. Soc. 120:11864-11873.

Yu, M., H. Davey, F. Svec, and J.M.J. Frechet. 2001. Monolithic porous polymer for on-chip solid-phase extraction and preconcentration prepared by photoinitiated in situ polymerization with a microfluidic device. Anal. Chem 73(21):5088-5096.

CHAPTER SIX

Functional Organic and Hybrid Materials

CHAPTER SUMMARY

The Panel on Functional Organic and Hybrid Materials addressed what its members expect to be defining general concepts that will emerge in the next two decades to fundamentally change the science and engineering of organic and hybrid materials. Many of these changes will be truly revolutionary. The panel predicts that organic materials of high and low molar mass will continue to increase their penetration of military materials applications for the foreseeable future because of the clear advantages they have in terms of functional flexibility, low weight, and facile processibility—all leading to economic gain over the life cycle. This prediction is based on an extrapolation of materials developments over the last 50 years.

The panel has identified a number of research opportunities, among them:

- Promotion of the convergence and integration of organic and Si electronics and other semiconductor and photonics into hybrid architectures;
- New synthetic strategies to produce high yields of selected polymers with completely defined chemical structures and with enhanced homogeneity and purity;
- Computer modeling and simulation, accessible to experimentalists, to optimize chemical and structure selection for specific functionalities (organic materials, especially macromolecular, that display high photovoltaic and thermoelectric figures of merit are particularly valuable for military applications);

- Use of organic materials to provide robust defenses against laser threats to personnel and equipment; and
- Novel catalyst systems to provide in situ defenses by neutralizing chemical and biological attack.

DoD investments in such areas will maximize the development of important novel organic materials with specific military applications. If these opportunities are pursued, the panel expects that:

- Modeling will become a routine first step in organic materials development.
- Synthesis and processing of organic materials will tend to converge.
- Polymers of high purity with totally controlled microstructure will become available, with important applicabilities.
- Aggregates of organic materials on the nanometer scale will yield new opportunities in material functionality.
- Combinations of low- and high-molar-mass organic molecules with inorganic materials will become widespread, offering unique functional advantages.

INTRODUCTION

The Panel on Functional and Organic Hybrid Materials believes that organic materials of low or high molar mass are destined to play a vastly increased role throughout the entire spectrum of military applications for the foreseeable future. By virtue of their functional flexibility, facile processibility, and intrinsic low weight—all of which contribute to an economic advantage over the application life cycle—their penetration into regimes hitherto held by metallic and other inorganic materials will continue at an unabated, and perhaps accelerated, pace.

In this chapter, the panel discusses what it believes to be the defining general concepts that will emerge in the next two decades to fundamentally change the science and engineering of organic and hybrid materials. Advances based on these materials are foreseen in electronic and photonic devices, eye protection against laser weapons, lightweight full-color displays, photovoltaic energy collectors, protection against chemical and biological agents, and many other areas. Many of these changes will be truly revolutionary. The chapter concludes with a discussion of R&D priorities that can help to bring about these revolutionary changes in the areas of greatest interest to the U.S. military.

DOD NEEDS FOR FUNCTIONAL ORGANIC AND HYBRID MATERIALS

Electronic Devices

The military will always have a need for low-cost expendable and long-term durable electronic devices. Though current silicon technology is viable for today's devices, there will be a need for molecular electronic devices in the future. Single electron-conducting molecules, including small clusters of metal atoms (hybrid systems), may be the basic technology for advanced electronic circuits and components in future small electronic devices (Lewin, 2001). Gimzewski (2000) projects that Si-based complementary metal oxide semiconductors (CMOS) will reach their limit in 10-20 years. Molecular electronics is expected to surpass this technology, providing that new fabrication methods and probes will allow individual or very small numbers of molecules to be connected to create actual devices.

The current capabilities of information technology (IT) are primarily associated with information processing and transmission. In the future it is expected that acquiring and acting on information will be a critical need, and current semiconductor technology is not projected to meet these challenges. Polymer electronics (organic electronic materials) may be the enabling technology for future IT needs.

Microelectronics today is already on the road to nanoelectronics, but there are exponentially increasing costs and diminishing returns associated with building new integrated circuit (IC) fabrication capabilities. This high infrastructure cost is limiting competition and innovation from smaller companies, while the technical miniaturization challenges (wiring, power dissipation, etc.) are increasing rapidly. Alternative technologies are required to advance IT needs in the next 20 years (Xu, 2000).

Current electronic systems have excellent information-processing speeds, but the hardware is difficult to reconfigure or rapidly evolve into more powerful systems. Molecular computers based on combinatorial syntheses of complex families of materials may be used to create new reconfiguration and evolution concepts. This could ultimately connect electronic processing speed with molecular design, structure, and flexibility (McCaskill and Wagler, 2000).

Current state-of-the-art molecular electronics technologies will have many challenges over the next 10-20 years, but they have the potential to deliver 1,000 times better performance in IT application areas than can be achieved with existing materials and systems (Wada et al., 2000).

Quantum effects associated with nanometer size dimensions are already considered in the design of microelectronic devices. Thus, it is possible that all organic molecular electronics will find either a significant niche apart from conventional CMOS systems or that hybrid products will be created using both technologies (Seabaugh and Mazumder, 1999).

Polymers, including composites and hybrid systems, are easy to process/fabricate, low-cost, lightweight, and flexible; they can have unique structural features and be made very durable. Thus, polymer electronic devices may become competitive with semiconductor and metal devices (Rughooputh and Rughooputh, 1999).

Miniaturization in electronic circuits, and ultimately devices, will reach the scale of atoms or small numbers of molecules (molecular electronics). There already have been many instances where single molecules have been embedded between electrodes and demonstrated basic digital electronic functions (Ellenbogen and Love, 2000).

Among the challenges still remaining are to theoretically design new materials using computational chemistry, synthesize these materials, and assemble and connect the molecular circuits and components to create practical devices (Joachim et al., 2000).

Photonics

In 2020 the military will need to effectively control the human-to-human, human-to-machine (weapon or network), and machine-to-machine interfaces. This means being able to rapidly sense or obtain large volumes (terabytes) of information, quickly analyze the information, and react accurately in time frames (nano- to picoseconds) that may not be possible today with current electronic and photonic devices. For these reasons it is projected that a number of advanced developments must be made in the area of photonics in order to design and produce devices and equipment that will process the massive amount of information that the military needs to be effective in future operations, both in peacetime and in conflict.

Optical Limiting Materials

Progress in compact laser systems made laser weapons possible, but it is a challenging task to protect sensors and eyes from laser light damage because there is a wide range of laser systems with different temporal and spectral characteristics. Two strategies can be employed to achieve sensor protection: all-optical switches and optical limiters. Both devices use nonlinear optical effects. The challenge will be to synthesize materials that exhibit large enough optical nonlinearity and chemical and photochemical stability. If such optical limiting materials can be synthesized, their impact on soldier protection will be enormous.

Organic Light-Emitting Materials

Low-cost, addressable, lightweight, full-color displays on flexible organic substrates with efficiencies in the 50-100 lumen/watt range or higher have substantial implications for information dispersion at the soldier level. Their development will require integration of several technologies. From a fabrication point of view, the printing technology that can in principle apply to polymer materials (less obviously to low-molecular-weight organics and probably not at all to inorganics) will give integrated all-macromolecule systems a competitive advantage. However, probable advances in other display methodologies make the equation less predictable. Other areas of military application are in solid state organic light-emitting diode (OLED) and polymer light-emitting diode (PLED) lighting, again taking advantage of simpler fabrication.

Polymeric materials also form the basis for polymer lasers. At present there have been many displays of stimulated emission in chromophores using optical pumping. Electrically pumped systems have been demonstrated from low-molecular-weight organics and are certainly imminent from macromolecules also. Many technical issues remain, including the architecture of cavity designs, but it is highly probable that electrically pumped polymer lasers will become available soon. The materials issues to be resolved deal with obtaining sufficiently high excitation densities by avoiding defect structures and morphologies. Again, host-guest polymer systems represent an attractive focus for research over the next 2 decades. The implications of having low-cost, efficient polymer lasers in information storage and retrieval systems are profound (McGhee and Heeger, 2000; Friend et al., 1999).

Molecular Magnetic Materials

It is anticipated that through 2020 the field of molecular magnetic materials will produce advances in lightweight motors and electric generators that incorporate organic- and polymer-based magnets for substantial weight savings. The field will continue to expand in terms of materials options, synthesis and processing choices, and new phenomena unique to the organic-, molecule-, and polymer-based architecture. It is expected that the magnetic ordering temperature for some examples of this class will exceed 600 K together with approximate thermal stability. It also is expected that the organic-, molecule-, and polymer-based magnets will be combined with organic- and polymer-based conducting, semiconducting, and photonic materials in integrated multifunctional smart materials. An example of such integration would be "spintronics" devices that are all-organic. New processing options, such as self-assembly of structures, will be commonplace.

Photorefractive Materials

The photorefractive effect has long been recognized to possess great potential for military applications (Günter and Huignard, 1998; Solymar et al., 1996), including high-capacity optical memories, dynamic hologram formations, massive interconnections, high-speed tunable filters, phase conjugation, real-time handling of large quantities of information, and real-time relay lines for phase-array antenna processing. Numerous device concepts using inorganic materials have been explored. However, only a handful evolved into military devices to date. Inorganic materials are difficult to prepare with defined composition (impurity levels) and are expensive. The successful demonstration of new devices based on organic photorefractive materials depends heavily on the emergence of new materials and processes.

Photovoltaics

Soldiers are equipped with sophisticated electronics and communications gear. Portable power for soldiers is the most immediate application for large-area flexible photovoltaics. Field stations and mobile armor units would also benefit from access to such devices. While the efficiencies of inorganic semiconductor-based photovoltaic devices are being improved, there is great appeal for an organic counterpart. Organic materials, either as low-molar-mass compounds on a flexible support or as polymers, are attractive due to the typical prospects of facile large-area fabrication, mechanical flexibility, potential very low cost, and the ability to tune optical properties to match the absorption characteristics of the solar spectrum.

Membranes

Three major military needs for membrane materials have been identified in the next 20 years: (1) soldier protection from chemical and biological agents, which will be the driver of development of "smart" membrane technologies. Such membranes ideally will protect the soldier while detecting and reporting the nature of the chem/bio agent, followed by decontamination and reactivation of the sensing elements; (2) membranes with high throughput and selectivity for water purification; and (3) membranes for power (especially portable) sources.

Metal Organic Catalysts

The principal areas important to metal organic catalysts for future materials for defense are smart materials, fuel conversion, and self-healing structures. In the case of smart materials, embedded catalysts are expected to act through a feedback loop as both sensors and actuators; an example is a metal organic that senses a biohazard and actuates a catalytic antidote.

For fuel conversion, the catalysts may be for fuel reforming, along with a fuel cell, or for producing nutritional substances on the battlefield. One pressing need is for electrocatalysts for direct methanol oxidation. For fuel cells that run on hydrogen, the difficulty is hydrogen storage. One solution would be to generate hydrogen on demand by direct oxidation of methanol, a fuel that is easier to store and transport. These electrocatalysts, which typically contain Pt and Ru, require a delicate balance in the metal-on-hydrous oxide structure (Long et al., 2000). More rugged materials that protect this structure are needed.

The third area is self-healing structures. Here the need is for catalysts embedded in textiles that can initiate polymerizations to repair tears and punctures (White et al., 2001).

SPECIFIC AREAS OF OPPORTUNITY

This section describes the challenges and opportunities presented if functional organic and hybrid materials are to be inserted into militarily important applications. In addition, a few crosscutting and high-risk opportunities for use of these materials are described.

Electronic Devices

The basic components needed to create electronic devices on a molecular scale are wires, switches, rectifiers, and transistors.

Molecular Wires

An organic solid [tetrathiafulvalene (TTF)–tetracyanoquinodimethane (TCNQ)] that exhibited metal conductivity below 59 K was reported some time ago (Heeger and Garito, 1972; Cowan and Wiygul, 1986). Later, Shirakawa et al. (1977) reported a doped polyacetylene material that had a room-temperature conductivity of 500 $(\text{ohm-cm})^{-1}$.

Work has continued in this area with primary attention to polythiophenes, polypyrroles, and other highly conjugated organic, heterorganic, and organometallic systems (Skotheim, 1986; Joachim et al., 2000). Certain modified polythiophene structures (Figure 6-1) have shown superconductivity properties at 2.5 K (Schön et al., 2001; Skotheim, 1986).

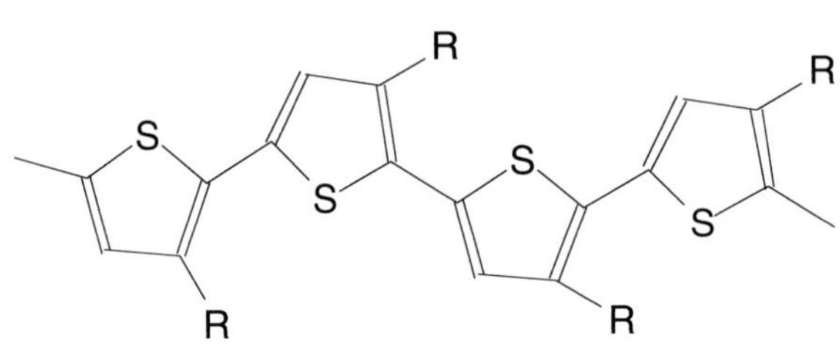

Poly (3-hexylthiophene)

R = hexyl alkyl group

FIGURE 6-1

Superconducting organic polymer.

Another approach to creating potential molecular wires is using σ bonds that are associated with polyorganosilane materials (Figure 6-2). These polymers exhibit semiconducting properties like photoconductivity, high hole drift mobility, and electroluminescence. They can be designed and fabricated into either rigid or flexible polymer chains that can be precisely connected to silicon surfaces (Fujiki, 1996). These types of polyorganosilane hybrid materials could find applications in bridging conventional silicon-based circuits with molecular organic-base building blocks or components (Fujiki, 1996).

Carbon nanotubes represent a special class of wires in that they are inherently conductive but their conductivity is propagated via tube-to-tube contact points or through substrate-tube-substrate connection geometries. Actually, the electronic structure of nanotubes can be metal-like or semiconducting, depending on the diameter of the tube and on the geometrical arrangement of the carbon atoms. This ability to have or create conducting or semiconducting properties has allowed several researchers to design and build nanotube diodes, T and Y junctions, and field-effect transistors (FETs) (Collins and Avouris, 2000; Lefebvre et al., 2000; Meyyappan and Srivastava, 2000; Dekker, 1999).

Switches

A simple molecular switch allows transport of electrons through a molecule while at the same time being able to disrupt the transport process in the molecule via conformational changes or other reversible reactions. For example, photochromic molecules can undergo a photo-induced intramolecular change in their molecular orbital structure that can favor or reduce electron transfer, depending upon the final molecular configuration

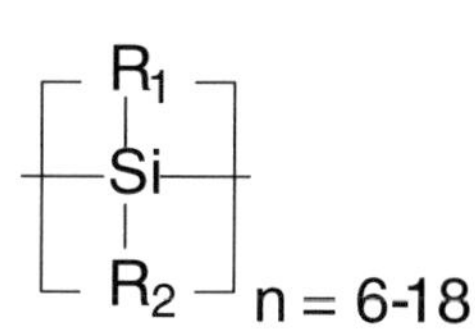

FIGURE 6-2

Potential molecular wire material that takes advantage of σ bonds in polyorganosilane materials. R_1 and R_2 are alkyl groups (branched and linear).

(Fraysse et al., 2000). Also, supermolecular structures, rotaxanes, are reported to have mechanical switching capabilities (Bissell et al., 1994).

Organic molecules that can be electronically switched on or off for extended periods might be used as the basic components of memory and logic devices. The molecules of choice are highly conjugated donor/acceptor functionalized phenylene ethynylene oligomers that are very sensitive to their local environments. When these molecules are constrained in a well-ordered monolayer stack of dodecanethiolate, their switching ability between conductive (on) and nonconductive (off) states under an electronic field is severely reduced. When their environment is changed to a less ordered structure, they can switch ability much more rapidly (Jacoby, 2001).

It has also been shown that when several thousands of these types of molecules are configured between gold electrodes, they can be switched between conductive and nonconductive states that allow data to be written, read, and erased just as in magnetic storage media (Reed, 1999a,b; Reed et al., 1997, 1998).

Rectifiers

Molecules that allow electrical conductivity in one direction through the molecule but not the other can be classified as rectifiers (Aviram and Ratner, 1994). An example of this type of molecule is γ-(n-hexadecyl) quinolinum tricyanoquinodimethane ($C_{16}H_{33}Q$-3CNQ) (Figure 6-3). This particular molecule has a high dipole moment (43 debyes) zwitterionic ground state (donor $^{+}$ –π bridge – acceptor^{-}) and a first excited state with lower polarity (donor ° – π bridge – acceptor°) and corresponding lower dipole moment (3 to 9 debyes). The ability to affect the flow of current in

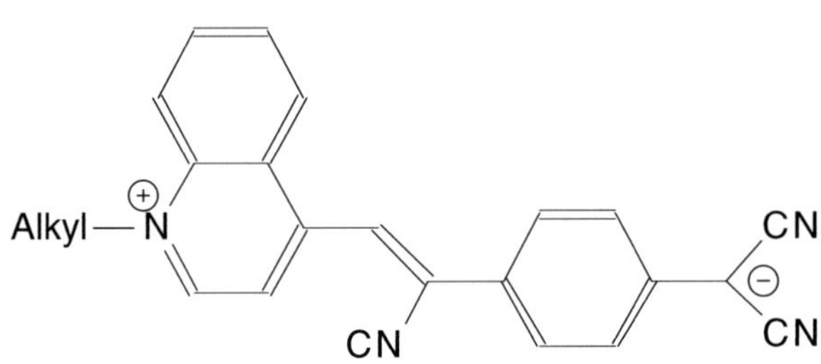

FIGURE 6-3

Molecular rectifier.

one direction and then in the opposite direction is a function not only of the difference in dipole moments of the molecule but also of how the molecule is configured between the electrodes of a test cell (Aviram and Ratner, 1974).

A Langmuir-Blodgett monolayer film of $C_{16}H_{33}Q$-3CNQ has been shown to rectify by intramolecular tunneling, while monolayers and multilayers tend to rectify as macroscopic film structures. These systems also show rectification between 105 K and 370 K, but there are concerns about voltage recycle capabilities and alignment stability of the film structures (Metzger et al., 1997).

Polymer Transistors

Transistors, as the major building blocks of any electronic circuit, should be the focus of interdisciplinary research teams in the future. Integrated circuits have been created using polymers in such conventional fabrication techniques as inkjet printing and microcontact printing (Garnier et al., 1994; Gelinck et al., 2000).

Figure 6-4 is a representation of a polymer field-effect transistor. The electrodes (gate, sources, and drain) were deposited by inkjet technology and the semiconducting or dielectric layers were created by conventional spincoating techniques. The polyimide channels were fabricated using photolithography and oxygen plasma processes on glass substrates, but polymer film materials, flexible or rigid, could be used as well. This particular all-organic polymer transistor had a high mobility of 0.02 cm^2/V-sec and its on-off current switching ratio was 10^5 (Sirringhaus et al., 2000).

Other recent developments in non-inorganic transistor technologies include the use of an organic semiconductor (pentacene) as the thin-film active layer transistor constructed on a glass substrate (Klauk et al., 1999). In the most recent molecular scale transistor (field-effect transistor), monolayers of 4,4′-biphenyldithiol were self-assembled on a gold substrate and then sandwiched under another gold top electrode. Two of these transistors (approximately 1,000 molecules) were used to create a "0" to "1"/ "1" to "0" input switch (Schön et al., 2001).

Electronic Circuits

It has now been established that most, if not all, of the individual basic molecular-scale building blocks that are required to create electronic circuits have been demonstrated. The next step is to create economically viable and highly reliable electronic circuits that are equivalent or superior

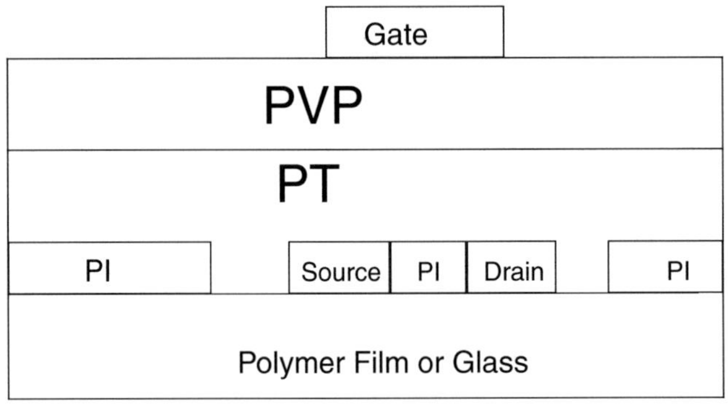

FIGURE 6-4

Representation of a polymer field-effect transistor. Gate, source, and drain = conducting polymer [poly(3,4-ethylenedioxythiophene) doped with polystyrene sulfonic acid]; PVP = polyvinyl phenol (dielectric); PT = poly(9,9-dioctylfluorene–co-bithiophene) (semiconducting); PI = polyimide (insulated). SOURCE: H. Sirringhaus, University of Cambridge, U.K., private communication, October 4, 2002.

to silicon technology but that operate on a molecular scale. Although a number of interesting organic/hybrid transistor components can be designed and put together and operate simple logic circuits, there still needs to be much research before we have a practical molecular processor that will replace semiconductor solid-state microelectronics (Drury et al., 1998; Reed, 1999a,b).

Alternative technologies like carbon nanotubes or molecular electronics vying to replace silicon technology for computing and data storage must meet certain criteria, such as high levels of integration ($>10^9$ transistors/circuit), high reproducibility (better than ±5 percent) and reliability (operating time >10 years), and very low cost (<1 microcent/transistor).

It is important to investigate new architecture alternatives to a CMOS-based conventional architecture that can take advantage of the unique electronic properties of the emerging nanomaterials. The use of genetic algorithms to generate novel architectures, evolvable hardware, and neuron model structures is now contemplated; such explorations may lead to computer systems based on nanomaterials and nanoelectronic devices. Since the future of nano or molecular electronics depends on the efficient commercial production of nanomaterials and it may be impossible to produce such materials without defects, it is also critical to invest in research on fault-tolerant architectures.

Opportunities and Challenges

The premise that molecular electronics will completely replace CMOS technology is probably not realistic. There is, however, a considerable amount of research in nanoscale CMOS, tunneling devices with III-V transistors, and single-electron devices that is projected to remain active throughout the 21st century. A current estimate of the time it will take for molecular electronics to become competitive with other technologies is between 2050 and 2100 (Seabaugh and Mazumder, 1999). Regardless of what time frame is projected for the replacement of CMOS electronic devices, the real need for miniaturization must be addressed. If molecular electronics is to provide a solution, partial or otherwise, for miniaturization of electronic devices, a number of major research and development hurdles must be cleared before high-performance, economic, and reliable devices can be produced at the molecular scale.

One of the first challenges for circuits composed of molecular wires, rectifiers, transistors, and so on is fabrication or assembly and connection of all the individual components. Photolithography chemistry, physics, and technology will have to be specifically adapted to the types of polymers and organic or organometallic building blocks that make up the circuitry. Though self-assembly of wires and rectifiers might be possible in the device fabrication process, the wires will still have to be attached to the termination points of the other components. Durability will also be important.

For all these reasons, the two major areas of molecular electronics research that need to be expanded between now and 2020 are

- Advanced computational design and then synthesis of durable molecular organic, polymer, and hybrid building blocks; and
- Advanced technologies for fabrication/assembly and component connection (Zhirnov and Herr, 2001; Kemp et al., 1998; Kelley et al., 1999).

A summary of some of the opportunities and challenges for the individual components required for molecular electronic circuits is given in Table 6-1.

TABLE 6-1 Summary of Where Research Is Needed to Develop Practical Molecular Electronics

Molecular Component	Opportunities	Challenges
Molecular wires	Viable method of connecting single molecules to a metal contact Use of self-assembly techniques to create molecular connections	Fabrication of single molecules between electrodes Maintaining molecular contact between electrodes during military operations Long-term durability of the electrical conductivity or resistance of the organic molecule
Rectifiers	A single monolayer film that exhibits current rectification by intramolecular tunneling	Voltage recycle capabilities over extended periods Maintaining alignment of the film structure over time
Molecular switches (bistable molecules or ions) and transistors	Ultrafast, nanoscale low-power circuits Billions of switches/transistors on a single chip Processing of increased information densities, especially if coupled to photonic devices (optical signal multiplexing and optical storage)	Control of the switch (limiting switching errors at the molecular level) Reversibility of the switch Readability at the molecular level Input/output attachments (nanowires or optical circuits) Heat generation and dissipation Fabrication using new techniques (scanning tunneling microscopy)

Photonics

Photonics can be defined as the science and technology associated with the manipulation of photons (light energies) for commercial and military applications. Photonics is concerned with the generation, transmission, switching or modification, amplification, and reception of light energy—which can be equated to information processing. Writing, sending, and reading information (optical communication) or manipulating information (optical computing) are just some of the areas that are important to the military, today and in 2020 (Bane and Bradley, 1999; Stix, 2001; McCarthy, 2001; Drollette, 2001).

The technological foundation of photonics associated with telecommunications is the use of devices both passive (optical fiber/waveguides, beam splitters/combiners, couplers, filters, and optical multiplexers) and active (light sources, optical switches or modulators, and detectors) (Figure 6-5). A future all-optical or hybrid (electronic/photonic) computer may have many of the same components, with the same functions as telecommunications equipment (Figure 6-6), but their size and time scale for operation will be considerably different (Eldada, 2001).

Light Generation

Semiconductor lasers (fixed and tunable), light-emitting diodes (LEDs), electroluminescent phosphors, and even miniature halogen or metal halide arc lamps are widely used today in telecommunications, sensor products, medical diagnostic applications, optical circuits, equipment inspection, and optical recording/reading devices. There is a major effort to commercialize organic LEDs for display applications. These light sources will be

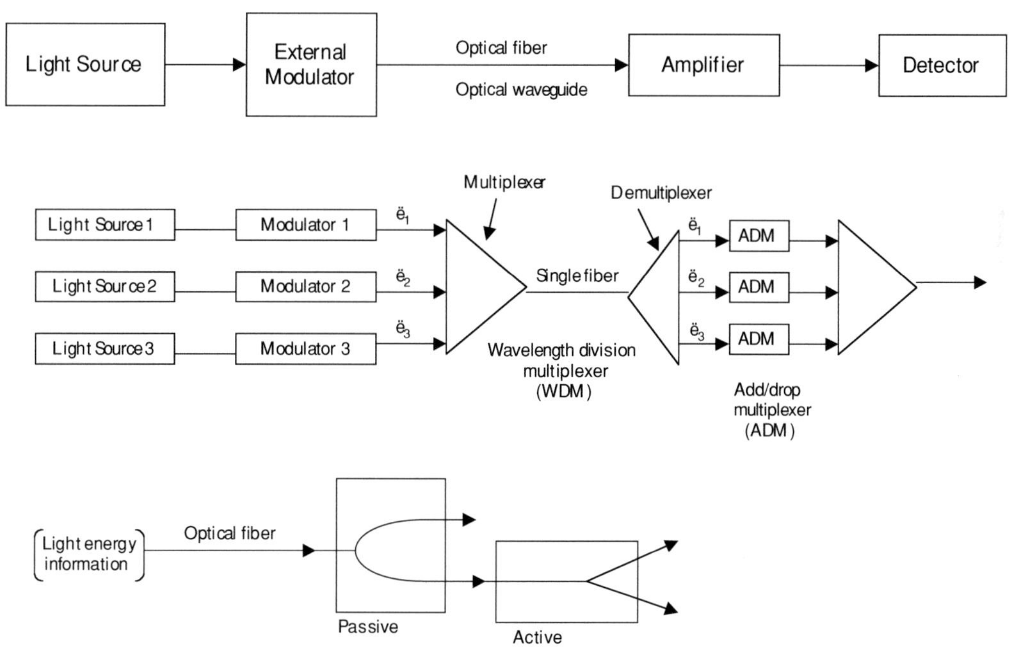

FIGURE 6-5

Photonic devices in the telecommunications industry. SOURCE: Reprinted by permission from Eldada (2001). Copyright 2001 by the International Society for Optical Engineering (SPIE), Bellingham,WA.

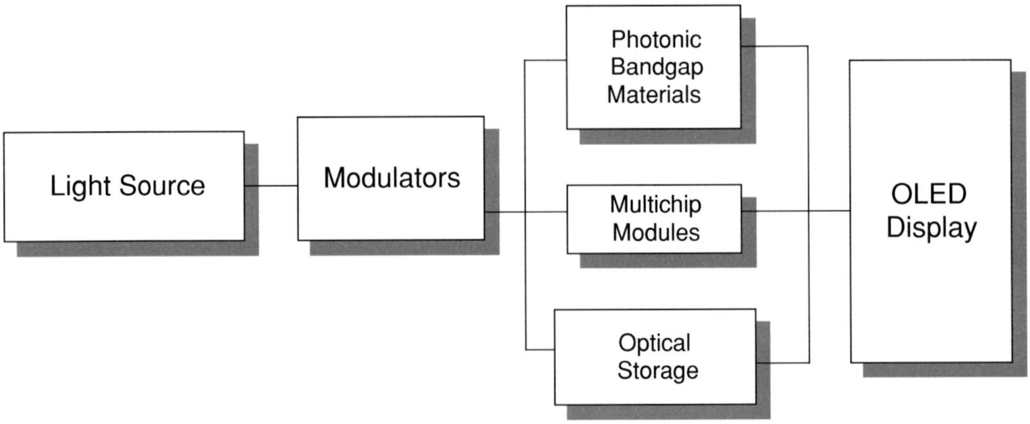

FIGURE 6-6

Potential photonic components for incorporation into all- or hybrid-optical computers. SOURCE: Reprinted by permission from Eldada (2001). Copyright 2001 by the International Society for Optical Engineering (SPIE), Bellingham, WA.

incorporated into devices that will be used by the military in the near future through 2020 (see discussion on OLEDs) (Giamundo, 2001; Krueger and Read, 2001).

Light Modulation (Optical Switching)

The telecommunications industry needs not only sources of bright, multiwavelengths of light energy but also a mechanism by which information is created within the light wave before it is transmitted to its destination. One method of creating information is to rapidly turn the light sources on or off (modulation) to create a digital signal. This method has limited capabilities (on/off rates of 2-3 GB/s) for rapid data generation. A better device technology keeps the light source output energy constant but very rapidly externally modulates (on/off rates of 10-40 or greater GB/s) the light signal after it is generated. These high-speed modulator devices can be fabricated out of either inorganic or organic-base materials; but the organic/polymer modulator devices are still in the research and early commercial development phase (Thomas et al., 2000a; Gogonea and Multhaupt, 1996).

High-speed, organic polymer-based, optical intensity modulators or switches operate on the principle of second-order molecular optical nonlinearity, or nonlinear polarization of highly unsaturated and aromatic compounds (chromophores) attached to a polymer backbone in an electric field. The electro-optic (EO) phenomena in the polymer come from chromophores that have a high second-order nonlinear optic effect and from the alignment of the chromophore molecules in an electric field. The two major factors that influence the EO effect of a chromophore are dipole moment (μ) and first molecular hyperpolarizability (β). Advances in molecular modeling software now make it possible to accurately calculate both μ and β for almost any type of chromophore; these calculations show excellent correlations with experimentally determined EO coefficients (Burland et al., 1994). All of the state-of-the-art EO chromophores have the basic structures shown in Figure 6-7.

It should be noted that all-optical processing (switching, computing) will require similar types of highly conjugated molecules, but these applications will depend upon the third-order, rather than the second-order, optical nonlinear effects of the molecule (Dalton et al., 1999a).

The most common device for either modulation or light-switching is a Mach-Zehnder interferometer that uses the polymer-chromophore as the waveguide and a special electrode configuration that activates the chromophore (changes its refractive index), causing an on/off disruption of the light propagating through the device (Figure 6-8) (Dalton et al., 1999a,b).

Light Transmission (Optical Fibers and Wave Guides)

The primary method for transmitting light over long distances (optical communications) is by using silica-base optical fibers, which have very low optical loss characteristics. This technology is continually being improved (Thomas et al., 2000a,b).

Polymeric optical fibers, the competing technology, have inherently high optical losses, but a number of synthetic approaches (deuteration or fluorocarbon modifications and graded index configurations) have improved their light transmission properties. Polymer optical fibers may never

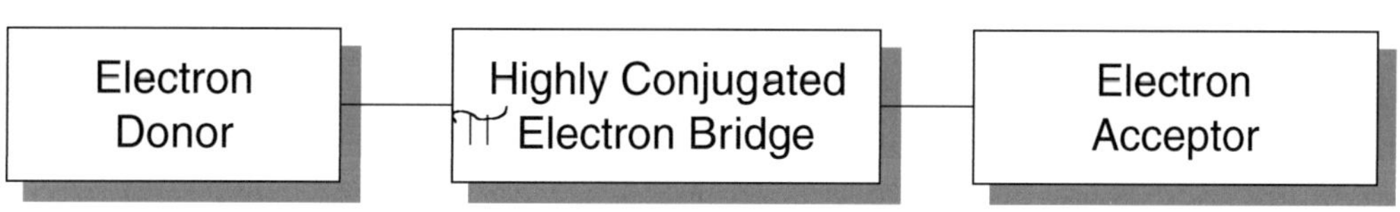

FIGURE 6-7

Basic structures of electro-optic chromophores.

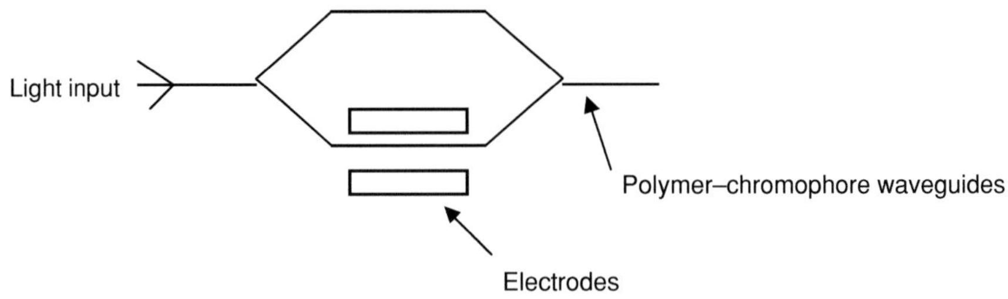

FIGURE 6-8

EO polymer-chromophore waveguide electro-optical modulator or switch. SOURCE: Reprinted by permission from Dalton et al. (1999a). Copyright 1999 by the American Chemical Society, Washington, DC.

replace silica fiber technologies for long-distance communication, but there are short-distance applications (transportation, sensors, product inspections) where polymer fibers have advantages (Kaino, 1992; Theis, 1992).

One area where polymers might compete with silica substrates is in flat planar waveguides, multichip modules, and other integrated optical circuit devices. Silica substrates for integrated optics are already well established; products like single beam splitters, thermo-optic switches, dense wavelength division multiplexers (Bragg gratings, arrayed waveguides and add/drop multiplexers [ADM]) are created using sophisticated integrated silicon-based optical circuits. These devices are highly reliable and can be scaled to a number of complex footprints, which helps maintain a cost structure that is economic for both the manufacturer and the user (Okamoto, 2000). However, polymers are easily processed or fabricated into products similar to silica-based products but that have lower costs and greater complexity of design and that are potentially easier to make connections with. They have achieved very high durability performance ratings in the laboratory (Driemeier, 1990; Wang et al., 1990).

Light Amplification

Optical amplifiers can be designed to enhance a signal directly inside a light source (semiconducting amplifiers) before feeding it into an optical circuit, or to preamplify a signal before it goes into a photodetector. In-line fiber optical amplifiers consist of silica glass fibers that have their cores doped with small amounts of erbium. A small perturbation (weak light signal) of the populated atomic energy levels of the f-shell of the erbium (Er) atoms stimulates the emission from these populated levels to cause amplification. The use of stimulated Raman scattering effects directly inside the optical fiber is another way to amplify a weak optical signal. It should be noted that erbium-doped fiber amplifiers (EDFAs) operate only at wavelengths between 1,500 and 1,600 mμ. Semiconducting amplifiers and Raman amplifiers can amplify a much wider range of telecommunications wavelengths (Thomas et al., 2000a).

There have been a number of examples of rare-earth complexes of Er^{3+}, Nd^{3+}, and Sm^{3+} being chelated, incorporated into polymer fibers or waveguides, and examined for their short-length amplification properties. If the stability of these hybrid metal/organic-polymer amplifiers can be improved, they might be used in a number of devices that contain complex (high-loss) optical circuits. Current technology is limited to a narrow range of energy wavelengths, but new developments in materials could expand the use of these types of amplifiers (Koeppen et al., 1997).

Photonic Bandgap Materials

The interaction of light with a material is strongly influenced by the dielectric constant or refractive index. Yablonovitch (2001) postulated that a material having the right pattern of defects or periodic structures should be able to trap and channel photons much as semiconductors control electrons. Semiconductors have crystalline structures that keep electrons in a specific energy range (electronic bandgap) that does not allow them to flow freely through the material. Doping a semiconductor with ions can alter or control the flow of electrons to create electronic circuits. If a pattern or lattice structure is created that has different optical properties, light can be trapped in an optical bandgap. This trapped light will not propagate unless a defect is created in the lattice that allows passage of the light in the materials (Yablonovitch, 2001).

A number of inorganic periodic structures have been fabricated that allow millimeter wavelengths of energy to be turned around a 90 degree corner, which is impossible for an optical fiber or waveguide. The advantage of these photonic bandgap materials is that they can control light in fully integrated optical circuits and thus become a basic foundation or building block for an optical computer (Levi, 1999).

Although the early work in photonic bandgap material is based on inorganic waveguide materials, it is possible to develop periodic dielectric structures based on incompatible blends of polymers. Block copolymers may be developed for photonic bandgap materials that can self-assemble into one-, two-, and three-dimensional periodic structures (Edrington et al., 2001).

Optical Storage

For a discussion of where polymer materials can be used in new optical storage media, see the section on photorefractive polymers.

Optical Computing

Very large scale integration is approaching certain limitations for silicon-base chips, and the increasing density of interconnections on chips could influence the development of all-optical or hybrid (electronic/optical) computers. The potential advantages for all-optical or hybrid computers derive from the fact that photons move faster (speed of light) than electrons, and that it is possible to make many optical interconnections for future massively parallel processing (MPP) computer systems. Optical interconnects can support high bandwidths (multiplexing capability), are reliable with low power consumption, and are immune to electromagnetic interference (EMI) effects while having low crosstalk capabilities. Guided wave technologies will be used to provide interconnection sites between chip-to-chip, module-to-module, and board-to-board components in all-optical or hybrid processors/computers (Karim and Awwal, 1992; Abdeldayem et al., 2000).

Photonic Materials: Summary

The photonic devices of the future will have to be configured so as to receive both electronic and photonic data from multiple sources (humans, sensors, machines). Once the information is received, it must be converted into all-optical signals and either analyzed in real time or stored optically for later information processing needs. It is also possible that the analyzed information packets will need to be transmitted rapidly to a receiver (human or machine) for further analysis, data confirmation, or storage. This final information packet might be used to activate another system as a final output of the analysis (Figure 6-9).

Table 6-2 contains a summary of where organic and polymeric materials need to be developed to create photonic devices that will have advantages over other technologies and meet military application needs in the future.

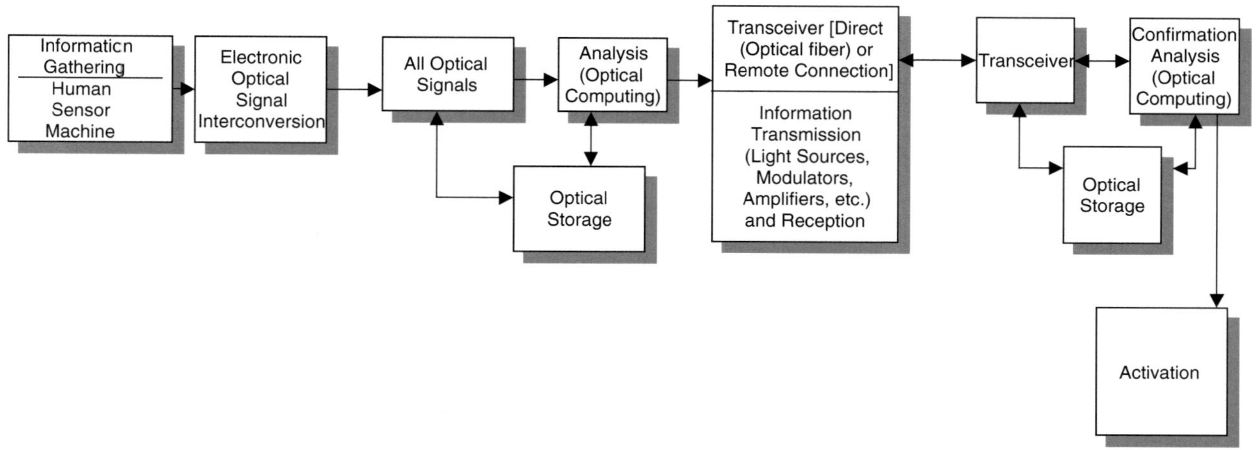

FIGURE 6-9

Potential military information gathering, analysis, and activation of another system.

TABLE 6-2 Summary of Where Organic and Polymeric Materials Might Be Used in Military Photonic Devices in 2020

Technology	Application	Advantages
Light sources such as organic light-emitting diodes (OLEDs)	Displays Low-power generation of multiple light signals	Flexible; can cover a wide range of viewing angles Multispectral signal processing
Polymer optical fibers, polymer waveguides, and polymer/hybrid optical amplifiers	Telecommunications Multiple-device connections Multichip modules Optical computer architecture and signal routing Signal enhancement	Not influenced by power surges Not subject to electrical shorts Not sensitive to electromagnetic interference No crosstalk between channels Light in weight Devices easy to reconfigure, connect, and integrate
Polymer photonic modulators and switches	RF signal distribution and control in space systems Rapid communication Optical computing applications Electrical-to-optical signal conversion Millimeter-wave signal generation Optical beam steering Detection of signals	High bandwidth (50 GHz or higher) Lightweight Not sensitive to electromagnetic interference High sensitivity at low cost Radiation resistant Low dispersion in the index of refraction between IR and millimeter-wave frequencies Easy to fabricate Low power consumption
Photonic bandgap materials, photorefractive polymers	Optical computers Optical data storage	Rapid computational processing speeds Able to process and analyze very large banks of diverse data

Optical Limiting Materials

Optical switches and optical limiters both use nonlinear optical (NLO) effects; the difference between the systems is illustrated in Figure 6-10. All-optical switches require a material that has a fluence threshold at which the material will turn completely opaque. Different schemes to achieve all-

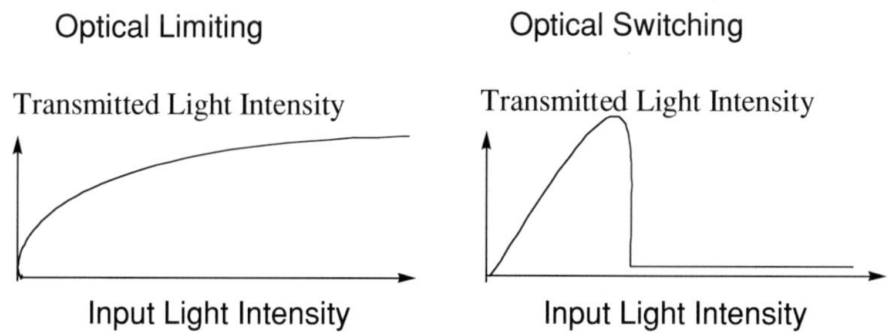

FIGURE 6-10

Schematic representation of optical limiting and switching.

optical switching using nonlinear optical effects have been extensively studied. Most of the device concepts use third-order NLO properties that are generally not yet large enough in almost every known material.

In optical limiters, light transmission is saturated when the light intensity reaches a certain level. Optical limiting can also be achieved by using different materials and mechanisms. Two-photon absorption and Kerr effects are two instantaneous nonlinear processes. Excited state absorption and free charge carrier absorption are accumulative nonlinear optical processes that require absorption and dissipation of light energy, and that can also be applied to optical limiting. The past several decades have seen studies of various optical limiting materials, such as inorganic semiconductor materials, organic or organometallic molecules, fullerenes, and nanometer semiconductor or metallic clusters (Tutt and Boggess, 1993; Spangler, 1999; Perry, 1997). Among these materials, carbon-60 and metallophthalocyanine complexes exhibit the best performance under certain conditions (Perry, 1997; Sun and Riggs, 1999; Miller et al., 1998; Tutt and Kost, 1992; Perry et al., 1996). Organic molecules with large two-photon absorption have also been synthesized and extensively investigated.

Organic Light-Emitting Materials

Significant electroluminescence (EL) in low-molecular-weight materials like alq_3 was demonstrated in the 1980s and in polymeric materials like poly phenylene vinylene (PPV) in the early 1990s. In both cases the prevailing mechanism is singlet decay from electron-hole recombination to excitons. The process is beset by nonradiative alternative decay mechanisms that severely limit quantum efficiencies. Nevertheless, as a result of intensive worldwide academic and industrial research, improved designs, materials, and fabrication processes have brought this area to the commercial exploratory stage in the relatively short period of one decade.

Technical issues that have received prominent attention include a priori design of high intrinsic quantum-efficient low-molecular-weight and polymeric light-emitting materials covering the full spectral range, blends and other nanostructured (self-assembled) materials of high efficiency, charge injection and mobility in organic materials, device architecture and ancillary drive designs, charge transport materials, rigid and flexible electrode substrate alternatives to indium-tin oxide (ITO), device mechanisms and lifetime improvements, and stimulated emission and lasing from EL materials.

Organic LEDs using both low molecular weight and polymeric lumiphores are now in full-scale production in the United States, Japan, and Europe. These are typically monochromatic displays, though full-color versions are on the immediate horizon, with applications in mobile telephone and small-scale (e.g. automobile) lighting. The efficiencies and lifetimes of these displays make production economically viable, though competition with new inorganic LEDs like III-nitride compounds and with alternative display technologies makes it possible that OLED display technology will secure only a niche footing.

In certain areas, OLED technology is likely to prevail by 2020. It is probable that architectures incorporating conformable, flexible substrates of any desired size will be created. These will incorporate active matrix full-color RGB pixels driven by hybrid or all-organic three-element circuitry. The economic advantages of such designs will largely center on low-cost fabrication, including roll-to-roll printing technologies.

The structure-property relationships required to optimize lumiphores include the challenge of predicting bandgaps and electron affinities by computational techniques. These will probably be developed by 2020, will permit color tuning (already available on a semi-empirical basis), and will be able to use more stable, higher work-function cathodes. To achieve high-efficiency white light emission, a multichromophore system must be devised either with conventional layer design or with nanostructured chromophore blends. The latter imply sophisticated morphological control and also offer the challenge of obtaining quantum-well or -dot structures of higher gain. Currently, three quarters of the excitonic energy is lost to triplet excitation states. By devising systems that facilitate intersystem crossing processes using heavy-atom or other effects, the theoretical quantum efficiencies can be significantly increased. Because many nonradiative decay paths arise because of impurities in the chromophore or from reactions occurring at interfaces, synthetic strategies to overcome these problems will be essential. Understanding of charge transfer processes at organic-metal or organic-organic interfaces is at a relatively primitive stage; increased attention to this area—theoretical, computational, and experimental—could have large payoffs.

Molecular Magnetic Materials

Magnetism arises from the quantum mechanical coupling (exchange interaction) between the spins located on atomic sites in a solid. Until 1985, all known magnets were based on the ordering of spins located on *d*-orbitals of transition metal ions and *f*-orbitals of rare-earth ions. The number of natural magnetic materials is limited; most are composed of transition metals (Fe, Co, Ni, and Gd), alloys, and oxides and prepared by high-temperature metallurgical or ceramic processes. In 1985 a revolutionary new platform technology was reported—organic-based magnets (Miller et al., 1985).

Though the first reported organic-based magnet (decamethylferrocenium tetracyanoethanide [DMeFc][TCNE]) ordered magnetically only below 5 K, this discovery offered an opportunity to use the methodology of synthetic organic chemistry to prepare magnets by design using low-temperature synthesis and processing options (Miller et al., 1988; Ovcharenko and Sagdeev, 1999).

The best candidates for high-temperature ferromagnetic materials are those containing magnetic transition metal ions. Purely organic ferromagnetic materials are difficult to prepare and the several reported systems are controversial. Prussian blue, $Fe_4^{III}[Fe^{II}(CN)_6]_3\ xH_2O$ (x = 14–16), a highly symmetrical octahedral coordination compound, is the simplest ferromagnetic molecular material; it undergoes a magnetic phase transition at 5.6 K. If the iron ions are replaced by other metal ions in variable ratios, very high critical spin transition temperatures (T_c) can be obtained. The highest T_c values are observed for $V_3[Cr(CN)_6]_2$ and $Cr_3[Cr(CN)_6]_2$ (Gadet et al., 1992; Mallah et al., 1992; Ferlay et al., 1995). The compound $V_{0.42}^{II}V_{0.58}^{III}[Cr(CN)_6]_{0.86}2.8H_2O$ is used to prepare a thermomagnetic switch and a device for absorbing solar energy. Another class of materials that exhibits high-temperature magnetic ordering is prepared from compounds of metals with the tetracyanoethylene radical anion (TCNE) with the composition of $M(TCNE)_2(CH_2Cl_2)_y$ (Zhang et al., 1998; Manriquez et al., 1991). Depending on the nature of the metal ions, a wide range of T_c is observed [T_c = 75 K, M = Mn^{II}; T_c = 75 K, M = Fe^{II}; T_c = 350 K, M = V^{II}].

Materials based on fullerenes and their derivatives are of interest. For example, tetrakis(dimethylamino)ethylene-fullerene shows ferromagnetism below 17 K and 3-aminophenyl-methano-fullerene-cobaltocene below 19 K (Narybetov et al., 2000; Allemand, 1991; Mrzel, 1998). More recently, a weak ferromagnetic signal was detected in polymeric C60 samples at temperatures up to about 227°C (Makarova et al., 2001). Although the origin of this magnetism is unclear, the results are exciting enough to justify further investigation.

As of 2001 more than a dozen new families of organic-, molecule-, and polymer-based magnets had been reported, with many of these families having numerous known members. The magnetic ordering temperature, T_c, was raised in 1991 to 400 K (125°C) with the preparation of $V(TCNE)_{\sim 2}$. In this case T_c is even above the decomposition temperature (350 K) (Manriquez et al., 1991).

By 2001 new processing choices had become available. For example, the $V(TCNE)_{\sim 2}$ with T_c of 400 K could initially be prepared only by reaction in solution. In 2000 it was reported that a low-temperature chemical vapor deposition (CVD) process operating at 40°C could be used to prepare films of $V(TCNE)_{\sim 2}$ that adhere to a wide variety of substrates, including glass, silicon, and Teflon (Pokhodnya et al., 2000). Further, these CVD-prepared films are substantially more air stable (unprotected samples remain magnetic in air for over two hours compared to less than a minute for the fine powders).

New phenomena particular to organic-, molecule-, and polymer-based magnets are appearing. These include organic semiconductors that are also magnets at temperatures considerably above room temperature. The magnetism of one class of molecule-based magnets could be increased or decreased in magnitude and ordering temperature by applying light of the correct wavelength. This phenomenon can occur below 20 K. A second class of organic-based magnets was reported to have photoinduced magnetism at temperatures as high as 75 K, nearly a fourfold increase in the temperature range for this phenomenon (Pejakovic et al., 2000). Yet another new phenomenon is the unusually strong frequency-dependent response to applied magnetic field for some systems.

Beyond ferromagnetic properties, paramagnetic molecules are also interesting and useful. An example is molecules that undergo photoinduced spin-crossover, a process in which the molecules transition from a low-spin to a high-spin state. More recently, Kahn and Martinez (1998) have described compounds that can undergo spin-crossover under thermal condition. They found that as the spin state changes, the color of the materials also changes. The information is stored as long as the temperature is kept within the hysteresis loop. Cooling down the materials erases the information. These materials can be used in display devices and are very interesting for such military applications as information storage and optical switches.

Molecular Magnetic Materials: Summary

Molecular magnetic materials can be expected to offer a range of properties that can be tuned by design and implemented by the methodology of synthetic chemistry. These tunable properties will include controlled mass density, mechanical flexibility, low-temperature processibility, high strength, solubility, low environmental contamination, compatibility with polymers for composites, biocompatibility, high magnetic susceptibilities, high magnetizations, high remanent magnetizations, low magnetic anisotropy, optical transparency, metallic, semiconducting or insulating conductivity, spin-polarized transport, and erasable photoinduced magnetism.

Given the pace of innovation in this field and the centrality of magnetic materials to many of today's technologies, we can expect that by 2020 there will be widespread use of organic- and polymer-based magnets both in current and in not-yet-proven technologies.

However, organic magnetic materials face strong competition from inorganic nanostructured magnetic materials, which have developed very rapidly in recent years. Nanoclusters made from transition metal like Co, Ni and alloys like FePt can be solubilized with organic ligands (Black et al., 2000; Sun et al., 2000). The size of these nanoparticles can be well controlled and they can self-assemble into three-dimensional superlattices (a typical example is the FePt nanoalloy particle). The self-assembly can be thermally transformed into ferromagnetic nanocrystal thin films that can be used in memory devices.

Despite the challenges and competition, organic magnetic materials possess unique features that other materials cannot match, among them versatility in structural modification to fine-tune properties, new properties like photo-induced spin crossover transitions, and ease in processing.

Photorefractive Materials

Organic photorefractive (PR) materials, which exhibit weak intermolecular interactions, are typically soft amorphous solids (Günter and Huignard, 1998; Solymar et al., 1996; Ducharme et al., 1991). In these materials, charge carriers are generated through dissociation of the tightly bonded excitions, not by interband ionization as in inorganics. The photogenerated carriers are transported under an electric field via intersite hopping. Due to the amorphous and disordered nature of the network, the depth of the traps in organic materials has a rather dispersed distribution. The EO response is provided by an individual molecular chromophore with an electronic origin (Moerner et al., 1997; Wang et al, 2000). The particular advantages of amorphous organic PR materials include a low dielectric constant, easy processing, and a high EO coefficient.

Although the PR effect is a complex phenomenon, a significant development of organic PR materials has been achieved in less than 10 years (Moerner et al., 1997; Wang et al, 2000). The advances in molecular engineering have produced numerous polymeric and molecular PR materials and three fundamentally different strategies for preparing PR polymers and organic materials: composite polymeric materials, fully functionalized polymers, and monolithic molecular materials.

Composite materials consist of polymer hosts (EO polymers, photoconductive polymers, or inert polymers) doped with different functional species as necessary for the PR effect. This approach has been successful for many PR systems. In general, there are four bases for composite PR polymers: (1) NLO polymers, (2) photoconducting polymers, (3) inert host polymers, and (4) liquid crystal matrices. In the first, a second-order NLO polymer is doped with charge sensitizers and charge-transporting molecules. In the second, a photoconducting polymer is doped with second-order NLO molecules and charge sensitizers. The third uses inert polymers as the host matrix for NLO chromophores and charge-generating and transporting species. The final class of PR materials is nematic liquid crystals doped with photosensitizers. In these materials, the major contribution to the photorefractive effect comes from the birefringence caused by reorientation of the liquid crystal molecules (Wiederrecht, 2001). Various polymers and molecular components are used in preparing PR composite materials, and numerous combinations of composites can be prepared. Several composite systems are outstanding in their PR performances

(Moerner et al., 1997; Wang et al., 2000). Overmodulation of the diffraction efficiency, large net optical gain, and large video frequency response time have been achieved.

A second option is fully functionalized polymers that contain all of the necessary functions for the PR effect in a single polymer chain (Wang et al., 2000). Different functional polymers have been reported to exhibit PR effects, including functional polyurethanes, functional conjugated polymers, functional polyimides, and conjugated polymers containing transition metal complexes. The most promising of these systems is the functional conjugated polymer containing metal porphyrin and phthalocyanine complexes. In this system, net optical gain and high diffraction efficiency exist. Unlike the composite materials, these functional polymers are thermodynamically stable and will not undergo phase separation.

More promising is a simple class of amorphous PR molecular materials exhibiting high PR performances. These materials show excellent processibility and transparency, long-term stability and durability, significant orientational enhancement, large net optical gain, and high diffraction efficiency.

Several prototype devices clearly show the advantages of using organic PR materials; these include devices for storage, retrieval, and subsequent erasure of digital data pages with a data density of 0.5 $Mbit/cm^2$ and for novelty filters and security systems.

To explore the practical application of organic PR materials, the ideal characteristics of materials for optical applications are high resolution, better energy sensitivity, broad wavelength sensitivity, real-time capabilities, good stability and compactness, and low cost. Although organic PR materials are among the most sensitive materials, numerous problems need to be solved. Improvements are needed to generate materials that respond faster, require lower applied voltage, and operate at the target wavelengths of telecommunication interest (1.3 and 1.5 μm).

Further improvements will depend on better understanding of the photochemical and photophysical details of the whole PR process: PR mechanisms in organic materials are still not well understood. These reasons may be due to the complexity of both the PR effect itself and PR materials. This is especially true for polymeric materials. A better mechanistic understanding will surely assist in the search for new PR materials with improved macroscopic properties. Funding of both device research and fundamental science is recommended.

Photovoltaics

The generation of electrical energy from sunlight is drawing much attention. Military interest derives primarily from the fact that such power can be generated and used (or stored) on demand. While the efficiencies of inorganic semiconductor-based photovoltaic devices are being improved, there is great appeal for an organic counterpart. Organic materials, either as low-molar-mass compounds on a flexible support or as polymers, are attractive because typically they offer the prospects of facile, large-area fabrication, mechanical flexibility, much lower cost, and fine-tuning of optical properties to match the absorption characteristics of the solar spectrum. To date, many limitations of organic materials have been recognized, and much fundamental research is still needed to address them.

The key component of a photovoltaic device is a material that can exhibit good photo-induced charge carrier generation with subsequent collection of carriers (Wallace et al., 2000). The basic steps are light absorption, charge separation, and charge collection. Light absorption can be quite efficient, although many conjugated polymer systems have rather large bandgaps and little absorption from the visible into the near-IR region.

A major problem with organic-based systems is limited charge separation of photo-induced carriers. The electron-hole (e-h) pairs created when light is absorbed are often bound as excitons with a relatively low dissociation tendency. Excitons are mobile, but their diffusion lengths are much smaller (<10 nm) than typical optical absorption lengths (>100 nm). It is desirable to efficiently dissociate excitons to allow holes and electrons to freely migrate to opposite electrodes. The third step, carrier transport, can be limited by low carrier mobilities. The challenge is organic materials that mitigate these limitations. Hybrid systems (e.g., silicon/organics) may offer some possibilities for improvement, although the ultimate goal is totally organic photovoltaic materials.

Strategies for enhancing carrier mobility include creating higher-purity, structurally perfect materials, perhaps as polycrystalline films with large crystal sizes, which can also help to increase the exciton diffusion length. Also, materials with bipolar (n and p) transport capability can help in exciton dissociation. Multicomponent organic materials might provide multiple absorption pathways and high interfacial areas between electron and hole transport regions that can provide a mechanism for exciton dissociation. A promising approach is the creation of molecular interfaces in a conjugated polymer, such as a substituted poly(p-phenylene vinylene) doped with C_{60}, which acts as an electron acceptor leading to e-h pair separation (Yu et al., 1995). A recent example is use of the discotic liquid crystal hexa-peri-hexabenzocoronene in combination with a perylene dye to produce thin films with vertically segregated perylene and hexabenzocoronene regions that have a large interfacial surface area that promotes e-h pair separation and subsequent transport. Incorporated into diode structures, these self-assembled films show photovoltaic response with external quantum efficiencies of more than 34 percent near 490 nm (Schmidt-Mende et al., 2001).

Finding the appropriate morphology and electrical and optical properties will depend heavily on processing conditions, so these must be carefully controlled. The self-assembly of liquid crystals is a promising approach, and there are opportunities for simultaneous synthesis and processing of organic-based photovoltaic devices. Toward that end, one proposal is to prepare stretch-oriented, conjugated polymer films with conductive polymer-dipolar-molecular materials acting as antennae and diodes to convert light to electric power (Marks, 1993).

There is another important practical consideration, namely, the stability of organic materials upon repeated exposure to light in the presence of oxygen and moisture. New materials with all chemical bonds having high dissociation enthalpies must be designed and synthesized, along with improved means of encapsulating devices. Attention must be given to electrodes attached to the organic components and how to minimize degradation at the electrode-organic interface. Finally, there is a need for new transparent electrodes that allow sunlight into the device with minimal absorption and with properties that make the electrodes amenable to large-area processing. While ceramics like tin-doped indium oxide are popular, there are opportunities for new organic transparent electrode materials. Here the bandgap would need to be small, so that doping-induced optical absorption would appear in the IR rather than the visible region of the spectrum.

Device processing considerations will also be important. Low-cost continuous coating processes are desirable.

The field of organic photovoltaics has been a modestly active research area. This panel believes it deserves much more attention if the goal of all-organic devices is to be realized in the next two decades. Like many others, this area is highly interdisciplinary. Steps must be taken to ensure that researchers interact productively in and between teams.

Membranes

Polymers are uniquely suited to be membrane materials because they can be processed as large-area films with properties that are readily tunable via chemical composition and processing. Applications include membrane-based separations (e.g., desalination of water by reverse osmosis) that represent an important class of operations in the chemical industry, ion-transporting films as solid electrolytes in power sources and organic LED devices, and protective clothing for combat personnel.

In separations applications, the penetrant material is allowed to pass through the membrane either by sieving through pores (for penetrants >2 nm in diameter) or by specific molecular interactions between the penetrant and the membrane material (typically for penetrants < ca. 2 nm). An example of the former is ultrafiltration. Reverse osmosis is an example of the latter: Water molecules interact with the membrane while solvated inorganic ions do not (solution-diffusion mechanism). Both mechanisms may be in play between these extremes (e.g., in nanofiltration). Ion-conducting membranes for power sources require electrical properties resulting from interconnectivity of ion-conducting paths, good dimensional stability, and, in fuel cells, the ability to suppress diffusion of anode and/or cathode reactants between the electrodes. High-temperature (120-180°C operation) membranes are being examined for hydrogen-based fuel cells to increase efficiency and minimize poisoning of catalyst sites by carbon monoxide from fuel reformate streams. In battery applications, membrane research is directed at structures that promote the facile transport of specific ions (e.g., Li^+) and on developing truly solid (e.g., nonplasticized) electrolytes having a high intrinsic free volume and hence high ion mobility.

Polymeric membranes will continue to present opportunities for significant growth in chemical and biochemical separations, sensors, gas and water purification, and electrolyte separators for batteries, fuel cells, and supercapacitors. There is increasing interest in highly selective membranes for various sensor applications where component separation before detection is essential. Water purification, especially desalination and wastewater treatment, will become even more important because of an expanding population and limited supplies of fresh water. Membrane materials that can self-heal (i.e., repair pin-holes or tears) will be particularly useful. Fundamental research on membranes will have spin-off potential in the medical arena in such areas as wound healing.

Membrane-based separation processes attempt to maximize selectivity while maintaining a high throughput. The medium is often asymmetric membranes, which have a thin skin affording selectivity and a porous backing layer that enhances throughput. Ideally, this two-layer arrangement would be largely unnecessary if the porosity of the membrane could be precisely controlled to molecular dimensions, affording true size-selective and, through functionalization of the pores, chemoselective separation. Indeed, inorganic materials like zeolites exhibit such characteristics; zeolite/polymer hybrids show promise as flexible membranes with good selectivity. Also, recent work on molecular imprinting of highly crosslinked polymers suggests strategies for preparing polymeric materials with similar capabilities that still retain many desirable features of polymers (flexible films and fibers, facile processing, easily tuned selectivity via organic functionality).

Block copolymers provide a useful platform by self-assembly of dissimilar block components into thermodynamically stable nanophases. As an example, in new proton-conducting block copolymer membranes for fuel cells, nanoscopic conducting domains spontaneously organize during processing. Block copolymers are also being considered as protective membranes for soldiers that can be selective about absorbing toxic agents yet be breathable for comfort due to the dual roles of the compositionally different, phase-separated domains.

A key challenge for defense interests will be producing breathable protective clothing for armed forces personnel that has embedded sensors and related electronics for the rapid assessment of environmental conditions and the binding and detoxification of various agents. Here, membranes will need to be smart (have sensing and reporting capability) in addition to having selectivity for toxic agents versus, for example, water. Also, materials that actually change barrier properties in response to a stimulus (chemical, electrical, magnetic) are desirable. Very thin membranes will require sensors and actuators of very small *z* dimension; here carbon nanotubes may play a role. Advances in materials processing and nanotechnology anticipated over the next two decades will have a direct and major impact in achieving multifunctional smart materials at reasonable cost.

Electrically and magnetically active polymers should also be considered, either as active membrane materials themselves or as embedded sensing and information-processing elements. For example, in electrically conducting polymers, the pore dimensions might be tuned using an electrical potential concomitant with oxidation/reduction chemistry and diffusion of counterions. Multicomponent polymer systems may also undergo dimensional and hence permeability changes as a result of field-induced modulation of microphase morphology.

This panel believes that membrane research will be a fertile area for DoD over the next two decades. Significant resources should be allocated to the fundamental science of modeling, synthesis, and processing of membrane materials, with an eye toward development of smart materials that can report on the local environment and change properties as needed, and can also repair defects. Attention should also be given to membranes capable of cleaning water with minimal energy expenditure, and to ion-transporting membranes for high-performance power generation (Koros et al., 1992).

Metal Organic Catalysts

A catalyst is a material that promotes a chemical reaction without becoming part of the product. By lowering the energy barrier for a chemical reaction, catalysts increase the reaction rate. In most chemical reactions, the goal is selectivity: chemoselectivity, stereoselectivity, chiral selectivity. To increase yield of the desired chemical product, the strategy is to find conditions that produce a controlled reaction, such as hydrogenation, hydroformylation, disproportionation, isomerization, carbonylation, or oxidation. In each case, a catalyst is used to promote the reaction and increase the yield.

Metal-organic catalysts are primarily metal alkyls, alkyls of Zr, Co, and Rh. Selectivity is a major factor in choosing a metal organic catalyst. For example, chiral selectivity allows the choice between two forms of a molecule that mirror each other; while one form may be beneficial or desirable, the mirror image may be harmful (Borman, 2001).

The focus of work on metal organic hybrids is homogeneous catalysis that can displace heterogeneous catalysts (Herrmann and Cornils, 1997).

The popularity of heterogeneous catalysts stems from the fact that they do not need to be recycled, and they have long service life, but homogeneous catalysts offer high activity on a metal-content basis. While it is difficult to recycle homogeneous catalysts, in some cases they can be recovered.

Metal organics contain ligands that keep the metal ion in a monatomic state of stereochemistry. The metal ion presents coordination sites for performing redox chemistry and coordination chemistry. Metal organic catalysts work on C-H, C-C, and C-F bonds.

Homogeneous catalysts should be studied in a step-by-step process to gain a mechanistic understanding of complexation, oxidative additions, reductive elimination, and insertion/migration mechanisms.

It is hard to draw a distinction between homogeneous and heterogeneous catalysts (Davies et al., 2001). The latter are designed to be completely removed from solution by filtration, but traces of heterogeneous catalyst may end up in the product. Presently, many of the encapsulating methods used with homogeneous catalysts render their behavior similar to that of heterogeneous catalysts. The newer methods of entrapping metal organics, such as sol-gel processing, reduce the leaching of the catalyst to the levels expected for heterogeneous catalysts (Blum et al., 1999). The rapid emergence of new encapsulating technologies, along with the high-throughput screening possible with combinatorial chemistry, has brought a wealth of new catalysts (Senkan, 2001).

Based on the successes of homogeneous metal organic catalysis, many new directions are being pursued. Two successes were hydroformylation, e.g., in the synthesis of Vitamin A, and carbonylation, e.g., in the conversion of methanol to acetic acid. There are active research programs looking into new ligands and new entrapment schemes, and there is interest in more sophisticated metal organics, such as metallocenes (Togni and Halterman, 1998), that are interesting for alkene polymerization. In addition, study of metallocenes should improve the fundamental understanding of mechanisms for designing better catalysts.

Among new directions for metal organic catalysts are the use of rare earth ions and bimetallic organic catalysts (Blum et al., 2000). In addition, nanoclusters of metals are a new development in precious metal organic catalysts. There are active efforts to produce metal organic catalysts supported on mesoporous materials and zeolites. Other new topics are organometallic chemistry in cells, and analogs to enzyme biocatalysis with metal organics.

Advances in synthesis have improved methodologies for designing metal organic catalysts. Work on polyoxometalates has led to stronger metal-ligand bonds (Schroden et al., 2001), and the capability of forming clusters. Sol-gel processes have been used to create better supports and high-surface-area materials (Avnir et al., 1998). Sol-gel processing also has been used to encapsulate catalysts to protect them from degradation.

New methods of complexation have given better ways to anchor catalysts. Silsesquioxanes are a class of materials that expand the types of support structures available (Zheng et al., 2001). With silsesquioxanes, mesoporous materials, and zeolites, other support structures are needed to prevent clustering and loss of activity. A new development in sol-gel encapsulation is the successful incorporation of an acid (molybdic acid) and a base (N-2-aminoethylamino-propylated silica) in the same silica matrix. This concept allows sequential acid- and base-catalyzed reactions to take place. Further extension of the concept is needed.

The perceived benefits of these hybrid organic-inorganic materials in the context of metal organic catalysts are that an inorganic matrix (1) improves thermal stability, (2) enhances chemical stability, (3) reduces air sensitivity, and (4) may increase selectivity through chromatographic behavior of the matrix porosity of the metal organic catalyst. Clearly, further R&D is required to optimize methods.

Transparent Electrodes/Organic Interfaces

A crosscutting issue in hybrid materials technology is transparent electrodes. In OLEDs, liquid crystal displays, photodetectors, solar cells, optical filters, electrical heating, anti-fogging devices and touch-screen sensors, the prevailing material is ITO. In passively driven displays, the anode and cathode are ITO on rigid substrates. In actively driven displays, the ITO is used in conjunction with thin-film transistor (TFT) arrays. The ITO layer is typically 100-500 nm thick, with >80 percent transmission in the visible light range. The ITO layer must be electrochemically compatible with metals like Al and maintain a low resistance (5-10 ohm/square). In addition, the ITO layer has to survive all processing of other layers in the device, including cleaning, patterning, UV/ozone exposure, and heating.

For all devices that rely on ITO, the present needs are for lower cost, fewer processing steps, reduced sensitivity to atmosphere to eliminate need for vacuum equipment, and better adhesion to flexible substrates to permit roll-to-roll processing. This list of improvements presents challenges that may or may not be met by ITO. Other transparent conductors, such as antimony-doped tin oxide (ATO), fluorine-doped tin oxide (FTO), and possibly ZnO, are being pursued. Parallel efforts need to be carried forward on both the processing and the chemistry sides to meet the requirements of planarization and adhesion.

Higher-Risk Developments

Examples of material developments that are as yet on the far horizon include the following:

Organic Thermoelectrics

These materials (especially the macromolecular ones) could provide extremely versatile power generation or cooling for a very wide range of military applications. Present (metallic) materials are expensive and difficult to fabricate and typically use heavy elements. Figures of merit (300 K) over Z = 1 have been demonstrated, but higher Zs are a material challenge (see Appendix D).

Organic Room-Temperature Superconductors

Oligomeric and polymeric regioregular thiophenes have demonstrated superconductivity in special circumstances at <5 K. As with ceramic/metallic superconductivity, what is needed is a material that will not only substantially increase T_c and current carrying capacity but also be easy to process. Such materials may become available by 2020; if so, they will have profound implications for myriad device and data-processing applications.

RESEARCH AND DEVELOPMENT PRIORITIES

In this chapter, the panel outlines its assessment of the opportunities that functional and organic hybrid materials offer for revolutionary new military capabilities by 2020. From this analysis, the panel has extracted five broad R&D priorities, discussed below, that are considered critical to the realization of these opportunities.

Convergence and Integration of Organic and Si (and Other Semiconductor) Electronics and Photonics in Hybrid Architectures

Opportunities will continue to arise for the use of polymers as components in hybrid materials systems, along with metals, ceramics, and electronic materials. Familiar composite materials (e.g., graphite-reinforced epoxy) will find additional applications, but significant potential is seen for new materials combination, such as organic electroactive materials and silicon for hybrid electronic and optical devices. Major issues will need to be addressed, such as how to match sometimes disparate properties like thermal expansion and optical absorption.

New Synthetic Strategies to Produce High Yields of Selected Polymers with Completely Defined Chemical Structures, Enhanced Homogeneity, and Purity

Synthesis and processing today are typically separate operations. This will change over the next two decades as opportunities emerge to carry out simultaneous synthesis and processing. This idea is not new (e.g., reaction injection molding or chemical vapor deposition), but it will become more widespread. Of particular interest will be polymers and small molecules that self-assemble into ordered molecular structures (e.g., liquid crystals) or morphological structures (e.g., block copolymers). Combinatorial synthesis of polymers will become more routine. The panel anticipates the emergence of combinatorial processing approaches to rapidly identify conditions for fabricating polymers to achieve maximum properties.

Polymerization techniques have limited opportunities to control the sequence of adding two or more monomers. Block copolymers are possible with successive addition of monomer charges to active chain ends, and alternating copolymers can be obtained under special circumstances. However, there currently is no viable means to prepare, for example, vinyl-type copolymers with sequence control (e.g., poly$[(\text{monomerA})_1-(\text{monomerB})_2]_n$). This is in stark contrast to peptide synthesis on ribosomes within biological cells, which employ a template to code for specific amino acids that are enzymatically linked. A major opportunity and challenge is thus sequence-controlled polymerization of a wide variety of monomers using systems that mimic the functions of ribosomes. Electrochemical polymer synthesis is an attractive option because properly patterned electrodes may simultaneously serve as solid templates for synthesis and as energy sources. The panel notes that other lessons from biology, such as self-assembly and development of hierarchical structures, will continue to be borrowed and built on.

Polymers are by nature complex materials, typically having a distribution of chain lengths, isomer content, degrees of orientation, and fractional crystallinity (if any). Thus, structure in polymeric materials can be hard to define compared to small molecules. This difference will gradually disappear over the next two decades with synthesis of long-chain molecules with greater compositional and structural precision. The implications will be significant: Very precise structure/property relationships will be possible, and through the integration of modeling, synthesis and processing, properties can be maximized.

Many common examples of nanoscale (one dimension of <100 nm) organic materials exist, including block copolymer films and collagen fibers that function as scaffolds for tissues and organs. However, there will be an increasing push to exploit the properties of individual molecules, or very small aggregates of molecules, for the next generation of electronic and optical devices. For example, carbon nanotubes and collections of only a few organic molecules are being studied as components of diodes, transistors, and memory elements. The ability to create well-defined organic molecular structures and manipulate them to form complex and functional arrangements will drive a revolution in information storage and processing, sensing, and communications.

Computer Modeling and Simulation, Accessible to Experimentalists, to Optimize Chemical and Structure Selection for Specific Functionalities

The creation of new organic materials will begin with broad evaluation of properties using high-level modeling and simulation to determine critical parameters (isomeric structure, molecular weight, degree of chain orientation) that influence a property of interest. In particular, modeling will be used to predict complex organization of functional low-molar-mass molecules and polymers, as is beginning to be done for the difficult problem of predicting protein-folding motifs. Modeling will also extend to synthesis routes to define the best approach, as well as to processing. Much guiding information will be in hand before any wet chemistry is done.

Organic Materials to Provide Robust Defenses Against Laser Threats to Personnel and Equipment

The challenge will be to synthesize materials that exhibit enough optical nonlinearity and chemical and photochemical stability. If such optical limiting materials can be synthesized, their impact on soldier protection will be enormous.

Catalyst Systems to Provide in Situ Defenses by Neutralizing Chemical and Biological Attack

Metal organic catalysts have a role to play in making materials multifunctional in the true sense of "smart" materials. In case of chemical or biological attack, embedded catalysts are expected to act as both sensors and actuators through a feedback loop. An example is a metal organic that senses a biohazard and actuates a catalytic antidote. However, new production methods, lower-cost catalysts, and new support structures will be necessary before these materials can realize their potential.

REFERENCES

Abdeldayem, H., D.O. Frazier, M.S. Paley, and W.K. Witherow. 2000. Recent advances in photonic devices for optical computing. Available online at <http://science.nasa.gov/headlines/images/nanosecond/thepaper.pdf>. Accessed February 21, 2002.

Allemand, P.M., K.C. Khemani, A. Koch, F. Wudlf, K. Holczer, S. Donovan, G. Gruner, and J.D. Thompson. 1991. Organic molecular soft ferromagnetism in a fullerene-C60. Science 253(5017):301-303.

Aviram, A., and M. Ratner. 1974. Molecular rectifiers. Chem. Phys. Lett. 29:277-283.

Avnir, D., L.C. Klein, D. Levy, U. Schubert, and A.B. Wojcik. 1998. Organo-silica sol-gel materials. Pp. 2317-2362 in The Chemistry of Organic Silicon Compounds, Vol. 2. Z. Rappoport and Y. Apeloig, eds. New York: Wiley & Sons.

Bane, P.W., and S.P. Bradley. 1999. The light at the end of the pipe. Scientific American 281(4):110-115.

Bissell, R.A., E. Cordova, A.E. Kaifer, and J.F. Stoddart. 1994. Chemically and electrochemically switchable molecular shuttle. Nature 369:133-137 [see also Borman, S. 1994. Reversible molecular switch synthesized. Chem. Eng. News, May 16, pp. 8-9, and Borman, S. 1999. Key step made toward molecular computing. Chem. Eng. News 77(29):11-12].

Black, C.T., C.B. Murray, R.L. Sandstrom, and S.H. Sun. 2000. Spin-dependent tunneling in self-assembled cobalt-nanocrystal superlattices. Science 290(5494):1131-1134.

Blum, J., D. Avnir, and H. Schumann. 1999. Sol-gel encapsulated transition-metal catalysts. Chemtech 29(2):32-38.

Blum, J., F. Gelman, R. Abu-Reziq, I. Miloslavski, H. Schumann, and D. Avnir. 2000. Sol-gel entrapped heteronuclear transition metal catalysts. Polyhedron 19(5):509-512.

Borman, S. 2001. William Who? Chem. Eng. News, 79(45):37-39.

Burland, D.M., R.D. Miller, and C.A. Walsh. 1994. Second-order nonlinearity in poled-polymer systems. Chem. Rev. 94(1):31-75.

Collins, P.G. and P. Avouris. 2000. Nanotubes for electronics. Scientific American (International Edition) 283(6):62-69.

Cowan, D.O. and F.M. Wiygul. 1986. The organic solid state. Chem. Eng. News, July 21, pp. 28-45.

Dalton, L., A. Harper, A. Ren, F. Wang, G. Todorova, J. Chen, C. Zhang, and M. Lee. 1999a. Polymeric electro-optic modulators: From chromophore design to integration with semiconductor very large scale integration electronics and silica fiber optics. Ind. Eng. Chem. Res. 38:8-33.

Dalton, L., W.H. Steier, B.H. Robinson, C. Zhang, A. Ren, S. Garner, A.T. Chen, T. Londergan, L. Irwin, B. Carlson, L. Fifield, G. Phelan, C. Kincaid, J. Amend, and A. Jen. 1999b. From molecules to opto-chips: Organic electro-optic materials. J. Mater. Chem. 9(9):1905-1920.

Davies, I.W., L. Matty, D.L. Hughes, and P.J. Reider. 2001. Are heterogeneous catalysts precursors to homogeneous catalysts? J. Am. Chem. Soc. 123(41):10139-10140.

Dekker, C. 1999. Carbon nanotubes as molecular quantum wires. Physics Today 52(5):22-28.

Driemeier, W. 1990. Bragg-effect grating couplers integrated in multicomponent polymeric waveguides. Optics Letters 15(13):725-727.

Drollette, D. 2001. Future-proofing defense. Photonics Spectra, May, pp. 104-112.

Drury, C.J., C.M.J. Mutsaers, C.M. Hart, M. Matters, and D.M. de Leeuw. 1998. Low-cost all-polymer integrated circuit. Applied Physics Letters 73(1):108-110 [see also Jacoby, M. 2001. Carbon nanotube computer circuits. Chem. Eng. News 79(36):9; Crone, B., A. Dodabalapur, Y.Y. Lin, R.W. Filas, Z. Bao, A. LaDuca, R. Sarpeshkar, H.E. Katz, and W. Li. 2000. Large-scale complementary integrated circuits based on organic transistors. Nature 403:521-523].

Ducharme, S., J.C. Scott, R.J. Twieg, and W.E. Moerner. 1991. Observation of the photorefractive effect in a polymer. Phys. Rev. Lett. 66(14):1846-1849.

Edrington, A.C., A.M. Urbas, P. DeRege, C.X. Chen, T.M. Swager, N. Hadjichristidis, M. Xenidou, L.J. Fetters, J.D. Joannopoulos, Y. Fink, and E.L. Thomas. 2001. Polymer-based photonic crystals. Adv. Mater. 13(6):421-425.

Eldada, L. 2001. Advances in telecom and datacom optical components. Opt. Eng. 40(7):1165-1178.

Ellenbogen, J.C., and J.C. Love. 2000. Architectures for molecular electronic computers: 1. Logic structures and an adder designed from molecular electronic diodes. Proceedings of the IEEE 88(3):386-426.

Ferlay, S., T. Mallah, R. Ouhes, P. Veillet, and M. Verdaguer. 1995. A room-temperature organometallic magnet based on prussian blue. Nature 378(6558):701-703.

Fraysse, S., C. Coudret, and J.-P. Launay. 2000. Synthesis and properties of dinuclear complexes with a photochromic bridge: An intervalence electron transfer switching "on and off." Eur. J. Inorg. Chem. 7:1581-1590.

Friend, R.H., R.W. Gymer, A.B. Holmes, J.H. Burroughes, R.N. Marks, C. Taliani, D.D.C. Bradley, D.A. Dos Santos, J.L. Bredas, M. Logdlund, and W.R. Salaneck. 1999. Electroluminescence in conjugated polymers. Nature 397(6715):121-128.

Fujiki, M. 1996. A correlation between global conformation of polysilane and UV absorption characteristics. J. Am. Chem. Soc. 118:7424-7425.

Gadet, V., T. Mallah, I. Castro, M. Verdaguer, and P. Veillet. 1992. High-T_c molecular-based magnets—A ferromagnetic bimetallic chromium(III) nickel(II) cyanide with T_c = 90-K. J. Am. Chem. Soc. 114(23):9213-9214.

Garnier, F., R. Hajlaoui, A. Yassar, and P. Srivastava. 1994. All-polymer field-effect transistor realized by printing techniques. Science 265:1684-1686.

Gelinck, G., T. Geuns, and D. de Leeuw. 2000. High-performance all-polymer integrated circuits. Appl. Phys. Lett. 77:1487-1489 [see also Levi, B.G. 2001. New printing technologies raise hopes for cheap plastic electronics. Physics Today 54(2):20-22].

Giamundo, S. 2001. A little enlightenment. Photonics Spectra, May, pp. 154-158.

Gimzewski, J.K. 2000. Nanomechanics and quantum mechanics of molecular systems. IEEE Seminar on Nanotechnology and Quantum Computing (Ref. No. 00/140), p. 6/1-2 [see also Gimzewski, J.K. 1998. Molecules, nanophysics and nanoelectronics. Physics World 1(6):29-33].

Gogonea, S.B., and G.R. Multhaupt. 1996. Nonlinear optical polymer electrets. IEEE Transactions on Dielectrics and Electrical Insulation 3(5):677-705.

Günter, P., and J.P. Huignard. 1998. Photorefractive Materials and Their Applications, Vol. 1, 2. P. Günter and J.P. Huignard, eds. New York: Springer.

Heeger, A.J., and A.F. Garito. 1972. Magnetic properties of conducting organic salts. American Institute of Physics Conference Proceedings, 18th Annual Conference on Magnetism and Magnetic Materials sponsored by IEEE. Am. Inst. Phys. 10(2):1476-1492.

Herrmann, W.A., and B. Cornils. 1997. Organometallic homogeneous catalysts—Quo vadis? Angew. Chem. Int. Ed. Engl. 36(10):1049-1067.

Jacoby, M. 2001. Molecular electronics. Chem. Eng. News 79(26):13 [see also Science 292:2303 (2001) and Appl. Phys. Lett. 78:3735 (2001); Tour, J.M. 2000. Molecular electronics. Synthesis and testing of components. Accounts of Chemical Research 33(11):791-804].

Joachim, C., J.K. Gimzewski, and A. Aviram. 2000. Electronics using hybrid molecular and mono molecular devices. Nature 408(6812):541-548.

Kahn, O., and C.J. Martinez. 1998. Spin-transition polymers: From molecular materials toward memory devices. Science 279(5347):44-48.

Kaino, T. 1992. Polymer optical fibers. Pp. 1-38 in Polymers for Lightwave and Integrated Optics, L.A. Hornak, ed. New York: Marcel Dekker.

Karim, M.A., and A.A.S. Awwal. 1992. Optical Computing: An Introduction, New York: Wiley.

Kelley, T.W., E.L. Granstrom, and C.D. Frisbie. 1999. Conducting probe atomic force microscopy: A characterization tool for molecular electronics. Advanced Materials 11(3):261-264.

Kemp, M., V. Mujica, A. Roitberg, and M.A. Ratner. 1998. Molecular wire interconnects: Chemical structural control, resonant tunneling and length dependence. VLSI Design 8(1-4):65-74.

Klauk, H., D.J. Gundlach, and T.N. Jackson. 1999. Fast organic thin film transistor circuits. IEEE Electron Device Letters 20(6):289-291.

Koeppen, C., S. Yamada, S. Jiang, G. Jiang, A.F. Garito, and L.R. Dalton. 1997. Rare-earth organic complexes for amplification in polymer optical fibers and waveguides. J. Opt. Soc. Am. B—Optical Physics 14(1):155-162.

Koros, W.J., M.R. Coleman, and D.R.B. Walker. 1992. Controlled permeability polymer membranes. Annu. Rev. Mater. Sci. 22:47-89.

Krueger, A., and I. Read. 2001. Chameleon lasers. Photonics Spectra 35(5):142-148.

Lefebvre, J., R.D. Antonov, M. Radosavljevic, J.F. Lynch, M. Llaguno, and A.T. Johnson. 2000. Single-wall carbon nanotube based devices. Carbon 38(11-12):1745-1749.

Levi, B.G. 1999. Visible progress made in three-dimensional photonic crystals. Physics Today 52(1):17-19.

Lewin, D.I. 2001. Beyond the wall: Computing with molecules. Computing in Science and Engineering 3(1):17-20.

Long, J.W., R.M. Stroud, K.E. Swider-Lyons, and D.R. Rolison. 2000. How to make electrocatalysts more active for direct methanol oxidation—avoid PtRu bimetallic alloys! J. Phys. Chem. B 104(42):9772-9776.

Makarova, T.L., B. Sundqvist, R. Hohne, P. Esquinazi, Y. Kopelvich, P. Scharff, V.A. Davydov, L.S. Kashevarova, and A.V. Rakhmanina. 2001. Magnetic carbon. Nature 413(6857):716-718.

Mallah, T., S. Thiebaut, M. Verdaguer, and P. Veillet. 1992. High-T_c molecular-based magnets—ferrimagnetic mixed-valence chromium(III)-chromium(II) cyanides with T_c at 240-Kelvin and 190-Kelvin. Science 262(5139):1554-1557.

Manriquez, J.M., G.T. Yee, R.S. McLean, A.J. Epstein, and J.S. Miller. 1991. A room-temperature molecular organic based magnet. Science 252(5011):1415-1417.

Marks, A. 1993. Light polarizing electrically conducting film, U.S. Patent 5,229,624 [July 20].

McCarthy, D.C. 2001. Tools to beat the bandwidth. Photonics Spectra 35(5):84-92.

McCaskill, J.S., and P. Wagler. 2000. From reconfigurability to evolution in construction systems: Spanning the electronic, microfluidic and biomolecular domains. Lecture Notes in Computer Science 1896:286-299.

McGhee, M.D., and A. Heeger. 2000. Semiconducting (conjugated) polymers as materials for solid-state lasers. Adv. Mater. 12(22):1655-1668.

Metzger, R.M., B. Chen, U. Hopfner, M.V. Lakshmikantham, D. Vuillaume, T. Kawai, X.L. Wu, H. Tachibana, T.V. Hughes, H. Sakurai, J.W. Baldwin, C. Hosch, M.P. Cava, L. Brehmer, and G.J. Ashwell. 1997. Unimolecular electrical rectification in hexadecylquinolinium tricyanoquinodimethanide. J. Am. Chem. Soc. 119(43):10455-10466 [see also Dagani, R. 2000. Taking baby steps to moletronics. Chem. Eng. News 78(1):22-26].

Meyyappan, M., and D. Srivastava. 2000. Carbon nanotubes. IEEE Potentials 19(3):16-18 [see also Menon, M., and D. Srivastava. 1998. Carbon nanotube based molecular electronic devices. Journal of Materials Research 13(9):2357-2361].

Miller, J.S., P.J. Krusic, A.J. Epstein, W.M. Reiff, and J.H. Zhang. 1985. Linear-chain ferromagnetic compounds—Recent progress. Mol. Cryst. Liq. Cryst. 120(1-4):27-34.

Miller, J.S., A.J. Epstein, and W.M. Reiff. 1988. Molecular ferromagnets. Account of Chemical Research 21(3):114-120.

Miller, M.J., A.G. Mott, and B.P. Ketchel. 1998. General optical limiting requirements. Pp. 24-29 in Proceedings on Nonlinear Optical Liquids for Power Limiting and Imaging, San Diego, CA, C.M. Lawson, ed. Bellingham, WA: SPIE.

Moerner, W.E., A. Grunnet-Jepsen, and C.L. Thompson. 1997. Photorefractive polymers. Annu. Rev. Mater. Sci. 27:585-623.

Mrzel, A., A. Omerzu, P. Umek, D. Mihailovic, Z. Jaglicic, and Z. Trontelj. 1998. Ferromagnetism in a cobaltocene-doped fullerenne derivative below 19 K due to unpaired spins only on fullerene molecules. Chem. Phys. Lett. 298(4-6):329-334.

Narybetov, B., A. Omerzu, V.V. Kabanov, M. Tokumoto, H. Kobayashi, and D. Mihailovic. 2000. Origin of ferromagnetic exchange interactions in a fullerene-organic compound. Nature 407(6806):883-885.

Okamoto, K. 2000. Fundamentals of Optical Wave Guides. New York: Academic Press.

Ovcharenko, V.I., and R.Z. Sagdeev. 1999. Molecular ferromagnets. Russian Chemical Reviews 68(5):345, and references therein.

Pejakovic, D.A.M., L. Jamie, J.S. Miller, and A.J. Epstein. 2000. Photoinduced magnetism, dynamics, and cluster glass behavior of a molecule-based magnet. Phys. Rev. Lett. 85:1994-1997.

Perry J.W. 1997. Organic and metal-containing reverse saturable absorbers for optical limiters. P. 813 in Nonlinear Optics of Organic Molecules and Polymers, H.S. Nalwa, and S. Miyata, eds. New York: CRC.

Perry J.W., K. Mansour, I-Y.S. Lee, X-L. Wu, P.V. Bedworth, C.T. Chen, D. Ng, S. Marder, P. Milles, T. Wada, M. Tian, and H. Sasbe. 1996. Organic optical limiter with a strong nonlinear absorptive response. Science 273(5281):1533-1536.

Pokhodnya, K.I., A.J. Epstein, and J.S. Miller. 2000. Thin-film V[TCNE]$_x$ magnets. Adv. Mater. 12:410-413.

Reed, M.A. 1999a. Progress in molecular scale devices and circuits. Pp. 104-107 in 57th Annual Device Research Conference Digest, June 28-30, 1999. Piscataway, NJ: IEEE.

Reed, M.A. 1999b. Molecular scale electronics. Proceedings of the IEEE 87(4):652-658 [see also Peercy, P.S. 2000. The drive to miniaturization. Nature 406:1023-1026; Miller, J.S. 1990. Molecular electronics. Advanced Materials 2:378; 2:495-497; 2:601-603].

Reed, M.A., C. Zhou, C.J. Muller, T.P. Burgin, and J.M. Tour. 1997. Conductance of a molecular junction. Science 278:252-254 [see also Reed, M.A., J. Chen, A.M. Rawlett, D.W. Price, and J.M. Tour. 2001. Molecular random access memory cell. Appl. Phys. Lett. 78(23):3735-3737].

Reed, M.A., C. Zhou, M.R. Deshpande, C.J. Muller, T.P. Burgin, L. Jones, and J.M. Tour. 1998. Molecular electronics: Science and technology, A. Aviram, and M. Ratner, eds. Ann. NY Acad. Sci. 852:133-144.

Rughooputh, S.D., and H.C.S. Rughooputh. 1999. Polymers for electronics and communications. Pp. 55-60 in 5th IEEE Africon, Electrotechnological Services for Africa. IEEE Region 8, eds. IEEE.

Schmidt-Mende, L., A. Fechtenkötter, K. Müllen, E. Moons, R.H. Friend, and J.D. MacKenzie. 2001. Self-organized discotic liquid crystals for high-efficiency organic photovoltaics. Science 293:1119-1122.

Schön, J.H., A. Dodabalapur, Z. Bao, C. Kloc, O. Schenker, and B. Batlogg. 2001. Gate-induced super conductivity in a solution-processed organic polymer film. Nature 410(6825):189-192 [see also Freemantle, M. 2001. Super conducting organic polymer. Chem. Eng. News 79(11):14].

Schroden, R.C., C.F. Blanford, B.J. Melde, B.J.S. Johnson, and A. Stein. 2001. Direct synthesis of ordered macroporous silica materials functionalized with polyoxometalate clusters. Chem. Mater. 13(3):1074-1081.

Seabaugh, A.C., and P. Mazumder. 1999. Scanning the issue. Special Issue on Quantum Devices and Their Applications. Proceedings of the IEEE 87(4):535-536.

Senkan, S. 2001. Combinatorial heterogeneous catalysis—A new path in an old field. Angew. Chem. Int. Ed. Engl. 40(2):312-329.

Shirakawa, H., E.J. Louis, A.G. Macdiarmid, C.K. Chiang, and A.J. Heeger. 1977. Synthesis of electrically conducing polymers: Halogen derivatives of polyacetylene $(CH)_x$. J. Chem. Soc. Chem. Commun. (16):578-580.

Sirringhaus, H., T. Kawase, R.H. Friend, T. Shimoda, M. Inbasekaran, W. Wu, and E.P. Woo. 2000. High-resolution inkjet printing of all-purpose transistor circuits. Science 290(5499):2123-2126.

Skotheim, T.A. 1986. Handbook of Conducting Polymers, Vol. 1 and Vol. 2, Skotheim, T.A., ed. New York: Marcel Dekker [see also Novák, P., K. Muller, K.S.V. Santhanam, and O. Haas. 1997. Electrochemically active polymers for rechargeable batteries. Chem. Rev. 97(1):207-281].

Solymar, L., D.J. Webb, and A. Grunnet-Jepsen. 1996. The Physics and Application of Photorefractive Materials, A. Hasegawa, M. Lapp, B.B. Snavely, H. Stark, A.C. Tam, and T. Wilson, eds. Oxford: Clarendon Press.

Spangler, C.W. 1999. Recent development in the design of organic materials for optical power limiting. J. Mater. Chem. 9(9):2013-2020, and references therein.

Stix, G. 2001. The triumph of the light. Scientific American 284(1):80.

Sun, S.H., C.B. Murray, D. Weller, L. Folks, and A. Moser. 2000. Monodisperse FePt nanoparticles and ferromagnetic FePt nanocrystal superlattices. Science 287(5460):1989-1992.

Sun Y.P., and J.E. Riggs. 1999. Organic and inorganic optical limiting materials: From fullerenes to nanoparticles. Int. Rev. Phys. Chem. 18(1):43-90, and references therein.

Theis, J. 1992. Polymer optical fibers in data communications and sensor applications. Pp. 39-69 in Polymers for Lightwave and Integrated Optics, L.A. Hornak, ed. New York: Marcel Dekker.

Thomas, G.A., D.A. Ackerman, P.R. Prucnal, and S.L. Cooper. 2000a. Physics in the whirlwind of optical communications. Physics Today 53(9):30-36.

Thomas, G.A., B.I. Shraiman, P.F. Glodis, and M.J. Stephen. 2000b. Towards the clarity limit in optical fibre. Nature 404(6775):262-264.

Togni, A., and R.L. Halterman, eds. 1998. Metallocenes: Synthesis, Reactivity, Applications. Weinheim and New York: Wiley-VCH.

Tutt, L.W., and T.F. Boggess. 1993. A review of optical limiting mechanisms and devices using organics, fullerenes, semiconductors and other materials. Prog. Quantum Electronics 17(4):299-338, and references therein.

Tutt, L.W., and A. Kost. 1992. Optical limiting performance of C-60 and C-70 solutions. Nature 356(6366):225-226.

Wada, Y., M. Tsukada, M. Fujihira, K. Matsushige, T. Ogawa, M. Haga, and S. Tanaka. 2000. Prospects and problems of single molecule information devices. Japanese Journal of Applied Physics, Part 1—Regular Papers, Short Notes & Review Papers 39(7A):3835-3849.

Wallace, G.G., P.C. Dastoor, D.L. Officer, and C.O. Too. 2000. Conjugated polymers: New materials for photovoltaics. Chemical Innovation 30:14-22.

Wang, M.R., R.T. Chen, G.J. Sonek, and T. Jannson. 1990. Wavelength-division multiplexing and demultiplexing on locally sensitized single-mode polymer microstructure waveguides. Optics Letters 15(7):363-365.

Wang, Q., L.M. Wang, J.J. Yu, and L.P. Yu. 2000. Fully functionalized photorefractive polymers incorporating transition metal phthalocyanine and porphyrin complexes as photosensitizers. Adv. Mater. 12(13):974-979.

White, S.R., N.R. Sottos, P.H. Geubelle, J.S. Moore, M.R. Kessler, S.R. Sriram, E.N. Brown, and S. Viswanathan. 2001. Autonomic healing of polymer composites. Nature 409(6822):794-797.

Wiederrecht, G.P. 2001. Dynamic holography in photorefractive liquid crystals. P. 319 in Optical Sensors and Switches, V. Ramamurthy and K.S. Schanze, eds. New York: Marcel Dekker.

Xu, J.M. 2000. Plastic electronics and future trends in microelectronics. Synthetic Metals 115:1-3.

Yablonovitch, E. 2001. Photonic crystals: Semicondutors of light. Scientific American 285(6):46-55.

Yu, G., J. Gao, J.C. Hummelen, F. Wudl, and A.J. Heeger. 1995. Polymer photovoltaic cells: Enhanced efficiencies via a network of internal donor-acceptor heterojunctions. Science 270:1789.

Yu, L., Q. Wang, M.-K. Ng, and L. Wang. 2001. Photorefractive effect in polymeric and molecular materials. P. 257 in Optical Sensors and Switches, V. Ramamurthy and K.S. Schanze, eds. New York: Marcel Dekker.

Zhang, J., J. Ensling, V. Ksenofontov, P. Gutlich, A.J. Epstein, and J.S. Miller. 1998. [M-II(tcne)(2)].xCH(2)Cl(2) (M = Mn, Fe, Co, Ni) Molecule-based magnets with T_c values above 100 K and coercive fields up to 6500 Oe. Angew. Chem. Int. Ed. Engl. 37(5):657-660.

Zheng, L., R.J. Farris, and E.B. Coughlin. 2001. Synthesis of polyethylene hybrid copolymers containing polyhedral oliogomeric silsesquioxane prepared with ring-opening metathesis copolymerization. J. Polymer Science: Part A: Polymer Chemistry 39(17):2920-2928.

Zhirnov, V.V., and D.J.C. Herr. 2001. New frontiers: Self-assembly and nanoelectronics. Computer 34(1):34-43.

CHAPTER SEVEN

Bioinspired and Bioderived Materials

CHAPTER SUMMARY

The Panel on Bioinspired and Bioderived Materials focused on how the integration of biology and the physical sciences could result in greatly improved, lightweight, multifunctional materials for DoD. Materials derived from biology, for example, biological molecules as the active element in sensors, and materials inspired by biology—for example, layered, hierarchical, abalone shell-like composites as lightweight, tough armor—were considered. The potential impact of applying biological paradigms to the development of materials to meet DoD requirements was reviewed in depth.

Biological systems have clearly shown that large numbers of molecules, structures, and systems in living organisms possess attractive materials properties that are beyond the reach of current nonbiological synthetic approaches. Many of these molecules, structures, systems, and natural fabrication processes could serve as the basis for synthetic materials with enhanced properties. The challenge of using living organisms as a model for materials for future defense needs lies in identifying defense applications and then understanding and manipulating the biological systems to solve them. However, the integration of biology and materials science is hampered by discipline-driven education and the historical separation of biology and materials science in academic, industrial, and military laboratories.

This chapter details specific DoD opportunities in the areas of structural materials, functional materials, materials for chemical and biological warfare, wound healing, and human performance enhancement. Each section contains a discussion of how an identified materials challenge might be met. The panel concluded that

- Biological toughening mechanisms offer a route to the next generation of lightweight, tough materials.
- Preservation of the biological function of biological molecules is a key driver for the next generation of biologically enabled devices.
- In vivo detection strategies to identify toxins and pathogens, including masked agents, may make it possible to detect a single agent molecule.

The panel identified the following priorities for DoD R&D investments:

- Improve fundamental understanding of the relationships between biological structure, properties, and evolution and materials design and synthesis.
- Increase communication of DoD material needs to biological and physical scientists.
- Pursue basic research into biological molecules, structures, systems, and processes to lay the groundwork for their use, or their use as models, in meeting DoD materials needs.
- Identify and produce biocompatible materials to enable in vivo implantable devices.
- Develop packaging technologies to preserve the biological function of biologically enabled devices.

In conclusion, biology can suggest directions for addressing many DoD needs in the next several decades. In some areas (e.g., improved battlefield medicine and identification or interdiction of biological warfare agents), solutions may be found in specific biological molecules, cells, or systems; in other cases (e.g., smart materials or lightweight structural materials), biology may point the way to improved strategies for material design and synthesis. Where biological molecules or cells are the active component of a device, the challenge to the materials community is not only incorporation of the sensing entity but, also, and perhaps more important, the preservation of biological function in a nonbiological environment. The common theme for all these technologies, research areas, and applications is to use biological paradigms for the solution of problems of materials design, materials synthesis, and systems assembly.

INTRODUCTION

Biological systems represent a successful strategy for the design of materials, the fabrication of parts and components, and the integration of parts into systems that meet complex performance criteria in a variety of environments. These evolutionarily engineered realities have emerged slowly. The possibility of incorporating the principles of evolutionary "strategy" into modern engineering and scientific practice is emerging in materials science and engineering.

This chapter is concerned with

- Identification of biologically derived materials for improved warfighting effectiveness,
- How applying biological paradigms will affect the development of materials meeting DoD needs, and
- Identification of biologically inspired materials for improved warfighting.

The current materials science paradigm is that a material's performance is uniquely related to its structure, and the structure of a material reflects the totality of its processing history. This insight allows processes and products to be developed based on scientific principle rather than purely empirical methodologies. Once the process-structure-property-performance database is established, a single material may be processed to satisfy a variety of applications. An example of this is the production of molecularly identical, morphologically distinct poly (ethylene terephthalate) fibers for applications as diverse as pliant textile yarns and stiff reinforcing cords for tires. Materials application in medicine has historically been highly empirical; seldom are materials primarily designed for biological or in vivo applications. The application of the methods of materials science to the understanding of biological materials should prove productive in finding new materials for DoD. An overview of the subject matter and disciplines treated in this chapter is shown in Figure 7-1.

The integration of biology and materials science is hampered by the fact that education is discipline-driven and by the historical separation of biology and material science in academic, industrial, and even military laboratories. This has created problems, among them incompatible usage of terminology, incompatible approaches to problem solving, and divergent cultural viewpoints, that can render interaction difficult.

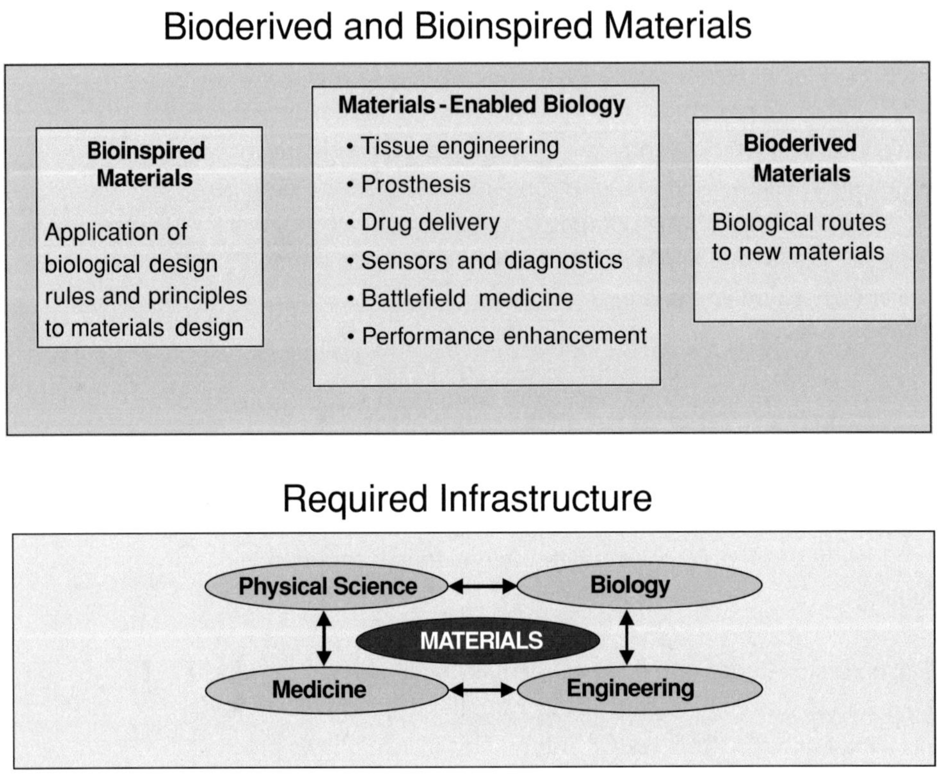

FIGURE 7-1

Schematic overview of subject matter and disciplines covered in this chapter.

The value of successfully integrating materials science and engineering with biology is difficult to estimate accurately but it represents at the least an opportunity to cross-fertilize and broaden each community's paradigms. Many of the examples cited below represent either the reality or the promise of this integration.

DOD NEEDS FOR BIOINSPIRED AND BIODERIVED MATERIALS

Mankind has been using biological materials for defense purposes for millennia: wooden staffs for spears or ax handles, vines for ropes, snake venom or plant extracts for biological weapons, shaped bone for tools and weapons. More applications are sure to come. As discussed below, multilayered abalone shell could point the way toward a new tough, lightweight vehicle armor, and spider silk fibers could become the basis for enhanced protection of personnel. Organic and biological electronics could lead to extremely light laptop computers and other devices. Insects may be seen as analogs of autonomous vehicles, equipped with composite smart armor featuring a variety of sensors and receptors. Conversely, a variety of biocompatible materials could be implanted (e.g., as hearing aids, lenses, artificial blood) to enhance soldier performance.

Billions of years of evolution have given mankind a wide variety of biological solutions to materials problems. The challenge lies in identifying relevant defense applications and understanding and manipulating the biological systems to solve them. It must be borne in mind, however, that the actual use of biological materials or materials that mimic biological systems lags far behind our enthusiasm for them. With a few notable exceptions, these biomolecular materials remain the materials of the future: despite an extraordinary rate of progress in the field, they are likely to be materials of the future for a number of years.

This does not mean that the field should be ignored by those focused on defense applications, awaiting a future when opportunities for applications might appear. Recent research has vastly increased our knowledge and understanding of biological materials and how their unique properties arise from their structure. New tools allow us to modify biological materials for our needs or to synthesize, de novo, materials that are based on biological principles. Nevertheless, there is a need for basic research to improve this understanding. More important, applications arise not from a linear progression of basic to applied research but from the vigilant and creative observation of basic research with an eye toward the unexpected, the unplanned, serendipitous breakthrough. Thus it is fair to say that basic research in all areas of biology must be pursued because there is no way to predict which discoveries in which fields will have important impact.

The greatest problem lies in linking those who discover intellectually interesting things to those who know of and are charged with meeting specific needs. It is not easy to find people who are both aware of current science and know the needs of the military. There is no doubt that there is much already in the basic science literature that could have defense applications but that has not yet been "discovered" by potential users. This will continue to be true as the pace of scientific discovery accelerates. To bridge this gap, basic researchers must be made aware of defense needs, so that they will notice potential applications when the underlying discovery is made; chance favors the prepared mind. At the same time, defense materials researchers must be closely linked to the basic research establishment so that they can extract potential applications from the flow of apparently nonuseful fundamental discoveries.

But the future lies in the biology. This chapter describes a wide variety of molecules, structures, processes, and concepts in biology that can reasonably be thought to have, at some point, a defense application. In some cases, research toward applications has made good progress. In many other cases, however, at this time we cannot even begin to develop those materials. In fact, in many cases we can list more barriers than routes to materials development. However, the great discoveries are those that even the most visionary cannot envision. Any concept that does not violate a law of thermodynamics should be regarded as possible. The shelves of military warehouses and consumer stores are stocked with valuable useful products whose function was regarded as impossible not many years ago.

> "Computers in the future may weigh no more than 1.5 tons."
> —*Popular Mechanics*, 1949

What, then, do living organisms do that might be exploited for defense applications? What do we know now about these molecules or processes and what do we need to understand to bridge the gap to application?

First and foremost, there is a clear need for better understanding of structure-property-function relationships in these materials. Second, it must be kept in mind that biological activities occur on the molecular level. Even the most complex of behaviors can (or will, once we understand them) be explained in terms of the presence or absence of specific molecules and the interactions among them. Further, it must be kept in mind that nature does not optimize biological materials and processes. Nature makes them just good enough—first, because perfection is not necessary for survival; second, because optimization of one process or structure might be detrimental to another; and third, because biological systems have evolved in the presence of a limited number of the 92 naturally occurring elements, within a narrow range of pH, temperatures, and pressures, and with the requirement that neither the product nor the intermediate molecules along the "production" path be toxic. Thus it is reasonable to argue that if we were to loosen the constraints, as we can when operating outside the organism, we might do better.

The greatest limitation to the mimicking of biological systems in materials and processes for defense applications lies in our insufficient understanding of these systems. One can argue that a more complete understanding should lead to many useful applications. On the other hand, it is equally clear that there are severe limitations on the use of biological molecules, structures, and processes. Biology does not in general thrive under extreme conditions. Biology limits itself by not making use of the vast majority of naturally occurring elements. Biological structures and systems are extremely complex, far more so than their synthetic analogues, and are therefore extremely difficult to manipulate—consider bone compared to titanium or even sophisticated alloys. These barriers could yield to our increased understanding but might also be avoided if the systems to be used were simplified, for example, in synthetic "cells" with artificial membranes enclosing volumes into which only a few well-defined molecules and processes have been incorporated to perform limited specific functions.

SPECIFIC AREAS OF OPPORTUNITY

Structural Materials

Bioinspired processing employs lessons from biology in creating synthetic analog composites. This approach is ideal for designing and fabricating nanostructured organic and organic/inorganic composites by mimicking the processes, structures, and properties of biological materials. Materials like bone, teeth, and shells are simultaneously hard, strong, and tough, with unique hierarchical structural motifs originating at the nanometer scale (Wainwright et al., 1982).

The structures of biological composites are hierarchically organized in discrete levels or scales. Virtually all biocomposite systems have at least one distinct structural feature at the molecular, nanoscopic, microscopic, and macroscopic scales. In the case of biological hard materials, nature grows hierarchically structured organic/inorganic composites in which soft materials (e.g., proteins, membranes, and fibers) organized at lengths of 1 to 100 nm are used as frameworks for the growth of specifically oriented and shaped inorganics (e.g., $CaCO_3$, SiO_2, Fe_3O_4, hydroxyapatite) with small unit cells (~1 nm). The high-modulus inorganic phase provides stiffness and the organic phase enhances toughness. Although the principle of hierarchical design has already been applied to synthetic composites (Lakes, 1993), techniques to reduce the smallest level of hierarchy to the submicron scale are not complete. Hierarchy at the nanometer scale leads to materials properties fundamentally different from those expected from simple rules for mixing the bulk properties of the constituents (Siegel, 1993).

Levels of structural organization are held together by specific interactions between components. For example, the structure of an abalone shell consists of layered plates of $CaCO_3$ (~200 nm) held together by a much thinner (<10 nm) "mortar" of organic template (Figure 7-2). Whatever the nature of the bonding between levels, adequate adhesion is required for the structural integrity of the system. Structurally organized organic surfaces induce growth of specifically oriented, dissimilar constituents catalytically or epitaxially. Highly interacting levels are organized into a hierarchical composite system designed to meet a complex spectrum of functional requirements. As composite systems increase in complexity, they perform better; the so-called intelligent materials and adaptive composite systems result from complex architectural arrangements. A hierarchical biocomposite is more than just a material from which larger objects can be built. It is a complete structural system where nested levels of structural hierarchy appear to yield improved dielectric and mechanical properties for particular functions.

The use of molecular biology and genetic engineering to produce materials has focused mainly on health care, resulting in, e.g., a large number of recombinant proteins for human therapeutics. Many of these products have been on the market for years, including insulin, human growth hormone, factor VIII, erythropoietin, and tissue plasminogen activator. Recombinant proteins for human health care have one common feature—high cost.

On the other hand, the biological synthesis of functional or structural materials using recombinant DNA technology has not progressed as rapidly as materials for human therapy. One of the main reasons is the lower value added for bioderived materials not targeted to the health care market. It is our belief that there is great potential for using biotechnology to produce biomaterials that could be well suited to DoD needs in 2020. Some of these biomaterials have already been demonstrated to be technically feasible. On the other hand, there are many challenges and barriers to be faced before such biomaterials become realities. In this section, the

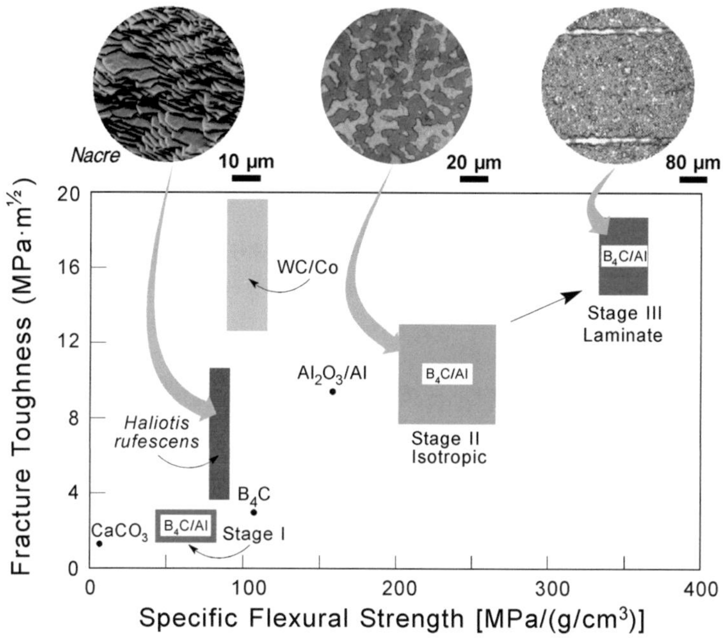

FIGURE 7-2

Mechanical properties of natural and synthetic materials. B_4C/Al composites—produced to mimic the interpenetrating and laminated structure of nacre—show significant improvement in properties over single-phase B_4C.

panel attempts to illustrate with examples as well as state some of the challenges that lie ahead.

Armor

An area that should be considered is how bioinspired structure might be explored for battlefield armor. For instance, the abalone shell has armor protection capabilities equal to or greater than those of existing materials on a strength-to-weight basis. When laminated hierarchical structures of biological systems (e.g., the nacre of abalone shell) are mimicked in microlaminated ceramic-metal, ceramic-organic, or organic-organic composites, significant improvements in composite mechanical properties are observed (Figure 7-2). Applying a simplified version of this layering to B_4C/Al (as well as SiC/Al or B_4C/polypropylene) composites significantly increases the mechanical properties. B_4C/Al composites are strengthened as a result of residual stresses with nanoscale modulations in the interpenetrating network of the ceramic and the metal phases.

Greenleaf Corporation at Saegertown, Pennsylvania, has manufactured B_4C/Al tiles for use in armor panels aboard C-130 and C-141 gunships. Processing of these ceramic/metal and ceramic/organic microlaminates is based on the concept of infiltrating a laminated scaffolding (e.g., ceramic) with a liquid (e.g., metal or organic polymer). The laminated composites produced by this approach are similar to nacre. However, although these accomplishments attest to the value of transferring lessons from biology and mimicking biological structures to create synthetic analogs, the smallest length scale in a complete system is still in the micron range due to the intrinsic limitation of the tape-casting method employed. On the other hand, the biosynthesis of inorganic biomaterials has not received the same magnitude of research funding as the human health care area. The panel believes that research is necessary to demonstrate the technical and economical feasibilities of such activities.

High Performance Fibers—Silk

It is well known that the natural product silk is one of the strongest and toughest fibers on a per unit weight basis. Of special importance is that silk performs the same whether it is subject to tensile or compressive forces, a property that differentiates silk from synthetic high-performance fibers. The amino acid sequences of many types of silk are now known. The production of recombinant silks based on parts of these sequences has been demonstrated in bacteria. Many companies and U.S. Army laboratories have conducted R&D on silk, among them Protein Polymer Technologies, Inc., DuPont, and the U.S. Army Natick Laboratories.

One of the major hurdles to the commercial development of recombinant silk is its cost. It is difficult today to compete on a cost basis for silk derived from the silk worm versus recombinant protein produced using a bacterial host. Recently, however, Lazaris et al. (2002) have been able to splice the genes for spider silk into cells from a variety of other organisms that, when grown in tissue culture, produce material that can be spun into silk threads. The groups plan to transfer the genes to goats that have been bred to produce the silk in their milk. However, only one of the two proteins of silk has yet been synthesized, and even that one is produced in a form that is shorter and weaker than the natural product. The barrier, as in most other examples discussed here, is biological. We do not yet fully understand the structure of the silk proteins, the mechanism by which the spider processes the proteins into fibers, and the techniques required to manipulate the extremely long pieces of DNA that code for the large silk proteins.

On the other hand, the integration of recombinant technology with materials science and engineering appears to be a natural partnership to improve the functionalities of recombinant biopolymers like silk over the silk produced naturally (Kaplan, 2002). For example, David Tirrell of the Department of Chemical Engineering at California Institute of Technology has proposed to replace the natural amino acids in native protein polymers using both chemistry and biotechnology to generate such materials with unique added functionalities. New types of polymers with well-defined selectable sequences and uniform composition are possible using these biological production methods. For example, one might see a modified silk with a significant higher strength-to-weight ratio that might be useful as material for the Armed Forces. The generation of highly complex biologically derived structures with new artificial properties and composition is possible.

Functional Materials

Soft Electronics

Recent advances have brought the electron mobilities for organic semiconductors within range of those for amorphous silicon (Heringdorf et al., 2001). This makes organic transistors practicable. Organic LEDs are also on the verge of full commercialization. The core device aspects of these advances are discussed in Chapter 6, but many processing and packaging questions still need to be resolved. DoD applications tend to be more demanding than consumer needs, with greater concern for vibration resistance, redundancy, damage tolerance, and environmental and corrosion resistance.

After the battery, much of the weight of portable electronics arises from the need for a rigid shell. Since silicon is inherently brittle and water-sensitive, for instance, there are high costs associated with the rigid packaging it needs for protection from moisture. The material must also be protected from the stresses associated with thermal expansion, acceleration, and vibration. Organic and/or biological electronics, deposited on flexible substrates, promise very considerable savings in weight. By integrating flexible photovoltaic cells, processors, memory, and display onto a substrate, it should be possible to make a laptop computer with the weight of a few sheets of paper.

A second advantage that would come from organic electronics, especially for the DoD, is that low-temperature printing processes would bring versatility. Small numbers of specialized circuits could be readily produced by combinations of inkjet and contact printing. Also, circuits could be printed onto external surfaces of equipment or onto clothing.

Molecular or polymeric devices are also more likely to provide a stable interface between electronic equipment and the human nervous system for performance enhancement, prosthetics, and health monitoring.

Biology offers insight into how self-assembled molecular-level electronic devices may be printed onto flexible substrates at low processing temperatures. To build devices that exploit the toughness of polymers and the simplicity of printing, we need to learn how to emulate biological processes and work with controlled disorder. We need a better understanding of the mechanisms of intermolecular electron or energy transfer, especially in disordered organic conductors. Much can be learned from their electron-transfer protein counterparts, such as the cytochromes. One approach to better understanding these transfer processes is to study single-molecule devices. Mark Reed of Yale University is studying such systems by addressing isolated organic molecules in a self-assembled monolayer using an AFM tip.[1] The work is focused on the molecule as a device; the greatest benefit will be in understanding of molecule-to-molecule electron transfer.

The processing methods being used for the most effective organic electronics depend on evaporation in vacuum of the active layers onto structures defined by photolithography. This is too expensive and slow. Printing methods are attractive but the resolution is not yet good enough. Methods that depend on self-assembly and selective deposition at boundaries where surface energy changes look most promising. Starting from studies of biomineralization, Aizenberg et al. (1998) have demonstrated mineral growth on surfaces with a linewidth of 10 microns (see Figure 7-3).

Another unmet need of molecular electronics is cost-effective flexible packaging systems. Organic LEDs, for instance, have aluminum electrodes that are very vulnerable to corrosion and oxidation and highly water-

[1]Reed, M., "Bioderived and Bioinspired Functional Materials," paper presented to the Panel on Bioderived and Bioinspired Materials of the Committee on Materials Research for Defense After Next, National Research Council, Irvine, CA, October 10, 2001.

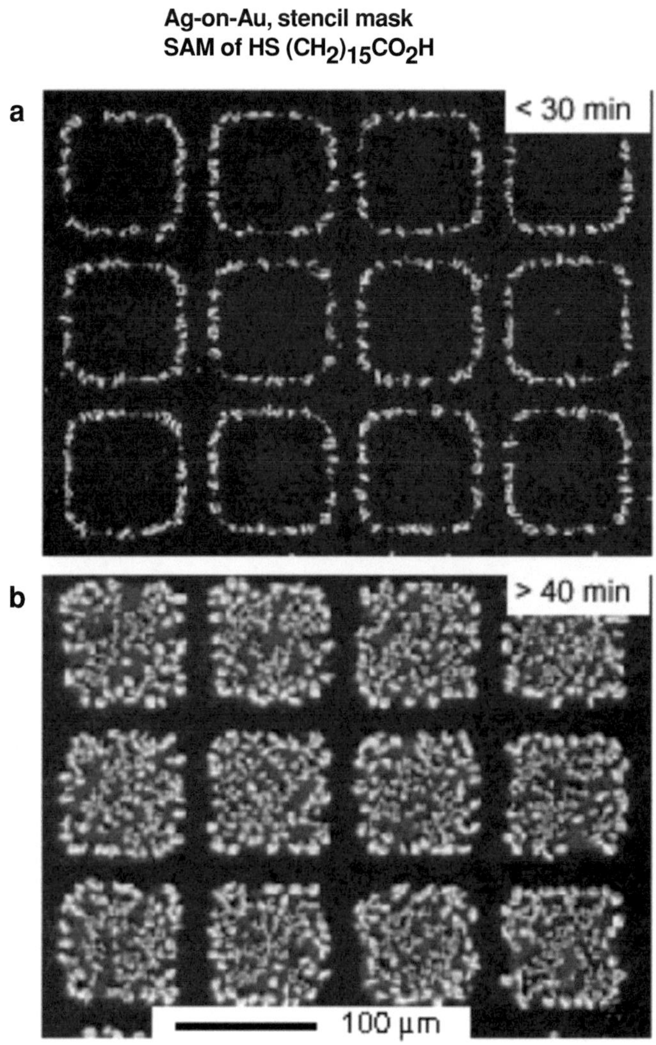

FIGURE 7-3

Calcite crystals grown on self-assembled monolayers on a patterned surface: (a) at short times, the deposition follows the outline of the pattern; (b) at longer times, the deposition reveals a resolution higher than the starting pattern. SOURCE: Reprinted by permission from J. Aizenberg et al. (1998). Copyright 1998 by Macmillan Magazines Limited, London.

sensitive electron conductors. They can be packaged as sealed modules between glass sheets. There is a need for an impermeable, transparent, and flexible packaging system. This could be a bioinspired multilayered organic/inorganic structure. Though no such material currently exists, biology may offer a way forward. Biological organisms came into being when the planet had a reducing atmosphere of methane and carbon dioxide. When photosynthesis caused oxygen levels to rise, special methods had to evolve to prevent destructive oxidation. Biomimetic active barriers that would deplete the water and oxygen that penetrate the system may provide part of the solution to this problem.

A challenging target for research in organic electronics is to create interfaces between cells, especially in the brain, and conventional electronics. This is being addressed by many groups, including those of Theodore Berger at the University of Southern California and Carolyn Bertozzi at Berkeley. Berger told the panel about implanting electrodes into living slices of tissue from the hippocampus.[2] Bertozzi discussed modification of cell surfaces to promote attachment to synthetic materials.[3] There is a host of applications awaiting better cell-machine communication. In current medical devices there is always an infection problem when wires or tubes penetrate the skin for extended periods. At present, interfaces, between nerve cells and silicon in vitro and between nerves and implanted devices, lead to cell death. There is evidence that conducting polymers may provide a more biocompatible interface. A stable interface could be used to allow amputees to control prostheses, for more direct man-machine communication, and to enhance such functions as sensing and memory.

Smart Materials/Sensors

Perhaps the most advanced smart materials at this time are sensors that translate their detection of defined targets into measurable optical, electrical, or mechanical changes. Many other organisms do this far more effectively than humans, demonstrating greater discrimination and sensitivity, sometimes approaching the detection of single molecules or photons. Dogs can distinguish individual humans by smell. Other organisms have exceptional senses of taste, touch, hearing, and sight. Biology thus offers

[2]Berger, T., "Memory Implants for Human Performance," paper presented to the Panel on Bioderived and Bioinspired Materials of the Committee on Materials Research for Defense After Next, National Research Council, Irvine, CA, October 10, 2001.

[3]Bertozzi, C., "Engineering Cell/Material Interfaces," paper presented to the Panel on Bioderived and Bioinspired Materials of the Committee on Materials Research for Defense After Next, National Research Council, Irvine, CA, October 10, 2001.

the possibility that individual soldiers, friend or foe, could be distinguished from a distance when sight and hearing are limited. Chemical or biological agents could also be detected before their concentrations reach toxic levels. In nature, a wide variety of mechanisms have evolved to achieve this kind of detection, any one of which could have an important military application. Lobsters and other crustaceans, for example, identify small differences in odor concentrations by flicking a pair of antennules, dragging them through the water to bring the chemosensory hairs that line them into contact with odor molecules (Koehl et al., 2001).

Quite different philosophies seem to guide sensor use in the biological and the synthetic worlds. Synthetic systems, such as the automobile engine, tend to rely on a few sensors, each highly reproducible and quite linear in its response to one variable. Biological sensing relies on large numbers of sensors that are very nonlinear, are time-dependent, and respond to multiple variables. Studies make it clear that a viper uses its heat sensors in combination with vision and chemical sensing on the tongue, making it very difficult to establish the role of each individual sense. In biology, the eyes of higher animals are perhaps the only stand-alone sensors, giving data that is separately interpreted. In many other animals eyes seem to act more as optical sensors than as cameras. Sensory data are usually interpreted only in the context of the full data stream for all the senses. Thus, cooling of the skin coupled with stimulation of the hairs means wind; without the hair sensor signal, it means extreme cold or evaporative cooling.

Mechanical sensors of animals can be separated into exteroreceptors, such as tactile sensors that sense changes in the environment, and proprioreceptors that allow control of the motion of the body. In mammals the proprioreceptors are embedded in the muscles, tendons, and joints. The most numerous sensors are the muscle spindles. These organs, which are 4-10 mm long, are incorporated in numbers from 50 to 500 per gram of muscle; the hand and neck have especially high densities. The entire human body has some millions of spindles. Muscle spindles are connected in parallel with the muscle and sense changes in length. In mammals, the Golgi tendon organs are located at the junction between the muscle and the tendon that connects it to bone. While the muscle spindles are sensitive to passive stretching of the muscle, the Golgi tendon organs are in series with the muscle and so sense the load on the muscle.

In biomimetic terms, insects can be seen as parallel to many types of equipment; there is an external protective shell around delicate internal organs. Arthropod cuticle contains integrated mechanoreceptors. One group of these is associated with sensing hairs; they act as exteroreceptors to provide information about the environment. A second group is associated with holes in the external skeleton, the cuticle. These proprioreceptors allow control of the motion of the animal by providing information about strains in the skeleton (Barth, 2002). Measurements of cuticular strain and response from spider legs subjected to applied loads gives a threshold sensitivity of about 50 microstrain (see Figure 7-4).

Insofar as insects can be viewed as prototype autonomous vehicles, there is clearly a particular DoD interest in this area. A consequence to be expected with the advent of printable electronics would be the embedding within equipment of many sensors that would provide continuous information about the device's state of operation and the surroundings. An initial stage in the process could be the development of smart skins for machinery. The availability of this information flow would then require control algorithms that allow an effective range of responses. There are many studies of embedded sensors for health monitoring and damage detection in composite materials and concrete structures. Fiber Bragg gratings are the most promising sensor (McKenzie et al., 2000), but there is not yet a simple way of integrating many sensors into a structure. There is a need for new miniature strain sensors with a simple readout. In insects, much mechanical sensing is through hairs that penetrate the cuticle. There is currently no way of making equivalent synthetic systems.

We need a much better understanding of biological molecular identification systems. Specific sensors, such as the moth antenna, are said to be able to detect a single airborne molecule of pheromone from the female. Specific biological sensing presumably operates as a lock-and-key, or more precisely as an induced-fit-through-conformational-change system where a set of intermolecular interactions are defined by the geometry of a receptor site and a complex target molecule. Recent advances in studies of cell-surface interactions have shown that short peptide sequences (such as RGD) can be specifically recognized. Likewise, supramolecular chemistry is now showing that several hydrogen bonds between two molecules of constrained geometry are sufficient for a strong and specific bonding interaction. If we wish to design molecular sensors, we need a better understanding of the relationship between molecular geometry and binding specificity. A critical issue is the preservation of the biological

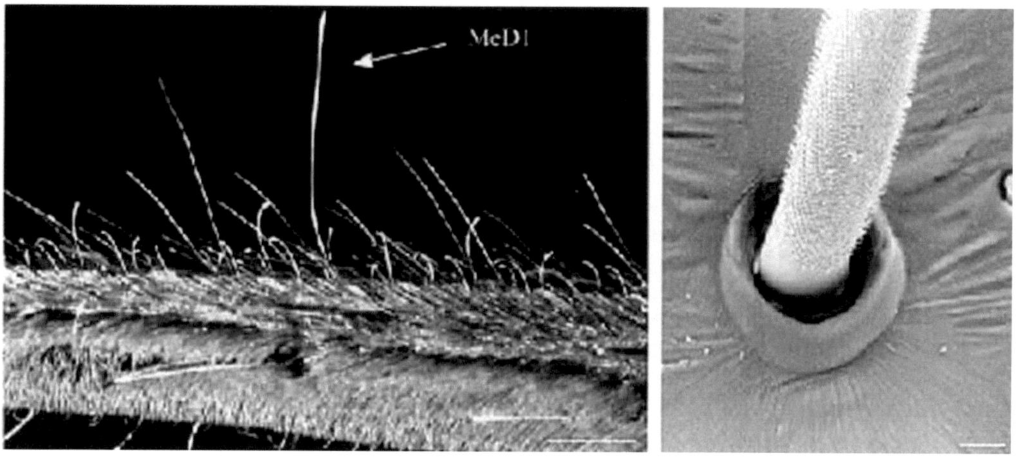

FIGURE 7-4

Tactile hairs on a spider leg. (Left) Metatarsus, scale bar 1 mm. (Right) Hair and cuticular socket, scale bar 10 microns. SOURCE: Reprinted by permission from J.T. Albert et al. (2001). Copyright 2001 by Springer-Verlag GmbH & Co. KG.

activity of cells and biological molecules used as sensing elements in a nonbiological environment.

Another area that is key to improved chemical sensor development is biomimetic amplification systems to parallel biological transduction mechanisms that facilitate detection of very small amounts of a chemical. There are at least two approaches. In the clotting system, binding of a few molecules of protein to a damaged surface sets off a cascade that results in a large amount of product. These systems can be intrinsically unstable unless they are backed with damping systems to eliminate subcritical responses. A second approach is to engineer a cell to produce fluorescent protein in response to a given stimulus. This case relies on activation of a gene, production of nucleic acid, and ribosomal protein synthesis, which will cause the response to be much slower. Whichever mechanism is used, amplification can be provided by immobilized cells in the sensor. This in turn requires that there be systems for maintaining cells for long periods of time before the sensor is used. It should be realized that this equipment will be stored, handled, and distributed more like food or medical supplies than like conventional equipment.

Power Sources

Batteries Batteries are a limiting factor in many devices. For example, laptop computer progress is stymied by a typical battery life of less than 8 hours for any acceptable battery weight. Thus, lighter power supplies, either fuel cells or batteries, are needed.

Batteries are composed of electrodes, electrolyte, case, and connections. Both the electrodes and electrolyte are multifunctional; there is a need for electrochemical activity, rapid transport of ions, mechanical stability to volume changes on cycling, and resistance to mechanical shock. The biological answer—hence the biomimetic solution—would be a complex composite structure integrating active components, conducting channels, and mechanical support. While composite structures have been considered for these tasks, they start from what can be made simply. As already discussed under structural materials, there is a need to design and build arbitrarily complex hierarchical structures, rather than simply characterizing those that appear from various types of processing. A research effort in this area should incorporate processing methods, modeling, and characterization tools at the nanometer, micron, and millimeter levels.

Recent years have seen intensive progress in new battery systems, especially secondary (rechargeable) lithium batteries. While there have been significant improvements, equipment makers still want far more energy than batteries can deliver at an acceptable weight. Batteries are characterized by both energy density and power density. Higher power density can almost certainly be achieved by going to multicell stacks of very thin layers with a hierarchical structure to minimize diffusion distances and internal resistance. Improved energy densities will either require fuel cells or wholly new systems.

Fuel Cells Fuel cells represent an attractive alternative to batteries. One could view the nervous system as an electrical system powered by glucose through oxidative phosphorylation. In the Krebs cycle, glucose is oxidized through a series of steps and the energy of each step is used to convert ADP (glycolysis and adenosine diphosphate) to ATP (adenosine triphosphate). ATP functions as the immediate energy source for many cellular processes, including muscle action. The process is efficient; about 70 percent of the energy available from glucose oxidation is converted into energy available from ATP.

There is a clear possibility of using similar biochemical or chemical routes to produce small-scale power. Glucose or sucrose would provide a cheap, convenient, and safe energy source. Table 7-1 shows that glucose has 70 percent of the energy density of methanol. Because the biological process uses electron-transfer proteins, a glucose-powered fuel cell can be based on enzymatic oxidation or on chemical catalysis. Prototype cells have been produced by a number of groups. Chen et al. (2001) recently described a fuel cell based on two carbon fibers, 2 cm long by 7 microns in diameter, that produced 0.6 microwatts, enough to power a simple CMOS device.

In both batteries and fuel cells, we have long recognized that the distance scale is the key to power density. Diffusion is fast over short distances, while thick electrolytes and electrodes require high ion mobility that compromises our structural choices. There is a clear need to build large structures with short effective length scales, just as a large mass of muscle has many interdigitated small units.

The Panel on Functional Organic and Hybrid Materials emphasizes the need for new methods to make polymers with well-defined sequences, analogous to proteins. Coupled to this must be methods to predict the structure and properties of the polymers, since the possible combinations are far too numerous to explore experimentally. Windle has considered the issues of multiscale modeling of polymers. His approach is to extract characteristic parameters from molecular modeling and use these to construct a simplified model at a higher level. In some cases the molecular

TABLE 7-1 Energy Density and Other Properties of Glucose, Compared with Chemicals More Commonly Considered for Producing Power

	$\Delta_c H°$ (kJmol^{-1})	$\Delta_c G°$ (kJmol^{-1})	n	E°	kJ/g	kJ/cm^3
Hydrogen	–285	–237	2	1.23	119	0.011
Methane	–890	–818	8	1.06	51	0.037
Methanol	–726	–702	6	1.21	22	17.37
Glucose	–2808	–2865	24	1.23	16	24.57
Octane	–5471	–5297	50	1.10	48	66.10

SOURCE: The Department of Chemistry of the University of Hong Kong. Available at <http://chem.hku.hk/~fuelcell/chem_fuel_cell.htm>. Accessed October 12, 2001.

information is reinserted once a higher level structure has been determined. Similar methods will be needed to design complex fine-scale composites that could mimic biological structure and function. Previous remarks about the need for flexible barrier materials for organic electronics also apply to batteries and fuel cells. Battery casings impose rigidity and are a significant source of weight.

Medical Applications

Chemical and Biological Warfare

Over 20 countries worldwide are developing chemical weapon (CW) capabilities, and at least 10 countries are working actively on biological weapon (BW) agents, and delivery vehicles. CW agents (e.g., phosgene, chlorine, chloropicrin, cholinesterase inhibitors) are relatively cheap and readily obtainable. They may be categorized as blistering agents or toxins against the nervous system, blood, and the respiratory functions. BW agents (bacteria, viruses, ricksettiae, genetically engineered microorganisms) are typically more potent than their CW counterparts, do not require massive stockpiles, are easy to conceal, and are produced with equipment that is commercially available with no legal restriction. Many agents are classified as midspectrum, between chemical and biological: Among them may be toxins from biological sources that adversely regulate pain, sleep, and blood pressure and act as physical and psychological incapacitants. Novel threats emerge from the engineering of biological entities to provide these with additional stability, resistance to antibiotics, or new delivery characteristics, or to make them nondetectable by engineered biosensors or the body's immune system. Additional threats derive from the release of toxic industrial chemicals (TICs) in the course of conventional warfare, as highlighted in the Yugoslavian theater.

The technology of sensors, even against conventional threats, is currently deficient, as highlighted by the experience in the Gulf War, and requires massive materials science and engineering-based advances to meet its C/BW mission.[4] Among reported deficiencies are limited standoff and liquid agent detection, potential for high false-alarm rates, weak single-individual biodetection technology, and slow response to central-

[4]Johnson-Winegar, A. Keynote address at the First Joint Conference on Point Detection for CBW Defense, Williamsburg, VA, October 23-27, 2000.

ized coordination. DoD research is directed to devices that will overcome such deficiencies, perhaps hand-held devices usable for detection, identification, quantification, and mapping of battlefield threats, as well as medical diagnosis and monitoring. These are expected to be all-agent capable and reprogrammable for detecting emerging threats. The fundamental scientific and technological advances required to accomplish these lofty objectives were discussed in *Opportunities in Biotechnology for Future Army Applications* (NRC, 2001), in particular as they relate to sensing and the battlefield environment (Chapter 3). One crucial conclusion emerging from the NRC report is that materials science and materials micro/nanotechnology have played a dominant role in contemporary breakthroughs in the detection and manipulation of biological molecules by way of sensors technology, micro/nanofluidics (e.g., Harrison et al., 1992; Desai et al., 1999), DNA chips (e.g., Christel et al., 1998), and protein chips (e.g., Bashir et al., 2001). It may well be expected that materials science and engineering will be on the forefront of the breakthroughs necessary for adapting and further evolving C/BW applications and civilian protection programs.

The 2001 NRC report also addresses biology-based electronics and computing technology (Chapter 4), biologically inspired materials (Chapter 5), biologically inspired solutions for logistics requirements, and novel technologies for the health and performance of warfighters (Chapter 7).

Wound Identification and Healing

Most battlefield casualties in conventional warfare result from uncontrolled bleeding. It is thus essential to develop methods, and the associated technology platforms, for (1) rapid triaging of battlefield wounds, (2) communication of the results to centralized and distributed battlefield coordination units, (3) summoning and deploying appropriate medical assistance, and (4) delivering medical treatment on-site while waiting for the medics to arrive (the required treatment includes hemostasis, antibacterial intervention, shock therapy, and possible reconstruction of damaged tissue).

The NRC (2001) has identified remote triaging of wounds, wound healing, and wounded tissue reconstructive engineering as primary R&D areas. Two of the five highest-priority areas identified pertain centrally to wound identification and treatment: self-replicating systems for wound healing; and shock therapeutics. An Army RFP was recently issued for the establishment of a major nanotechnology center with the primary objective of integrating wound triage and treatment technology directly into the warfighters' uniforms.[5] It said:

> Regardless of the injury mechanism (e.g., infection, trauma, surgery), several critical factors are common to the wound-healing process. These factors include an adequate blood supply to the healing tissue, resolution of associated infections, infiltration of the wound site by inflammatory cells followed by mesenchymal cells, and finally the deposition of neoconnective tissues and epithelial tissues.

An adequate blood supply to injured tissue has long been recognized as vital to healing. Cupping, the practice of applying a cup heated by a flame over the site of injury, was used for centuries to ensure blood flow to topical wounds. Today, angiogenic factors can be delivered to the sites of injury to stimulate the formation of new blood vessels.

Hyperbaric chambers have been devised to increase oxygen concentration for cells at the site of injury and thereby increase their viability and rate of proliferation. Acupuncture, massage therapy, and a variety of poultices have been used to create the optimal wound-healing environment, especially for recalcitrant nonhealing wounds.

Wounds have been categorized by their severity, depth, and chronicity. Each category has its own standards of care. However, the principles of cleanliness, wound covering, tissue apposition, and protection from physical trauma while tissues return to their normal physiologic state apply to all wounds.

A variety of coverings are used for acute and chronic wounds. Dressings range from totally occlusive dressings that do not allow fluid (and allow little gas) to pass from the wound to the outer environment, to partially occlusive or nonocclusive dressings that remain permeable to both fluids and gases. Dressings may or may not carry antiseptic or antibiotic compounds. In general, wound coverings for acute traumatic wounds are adequate for treating infections and protecting wounds from further injury. However, there is a pressing need for wound coverings that simultaneously provide, protect, and deliver a stimulus for healing. Stimulation for healing is especially important for large injuries when "space" must be "filled": In natural healing, large pockets at sites of injury are filled with fluid (usually plasma, blood, or both), which then creates a barrier to rapid healing. Therefore, dressings that not only cover the wound but also

[5]Badylak, S.F., Department of Biomedical Engineering, Purdue University, private communication, June 28, 2000 (as cited in NRC, 2001).

stimulate the formation of new blood vessels and the deposition of connective tissue would greatly improve wound care. Bioerodable polymeric biomaterials specifically designed for biological and mechanical responses in vivo, such as those developed in the laboratories of R. Langer and others (Wise, 2000), provide the basis for these needed improvements but their introduction is hampered by quality issues and cost.

There is an inevitable gap between in vitro phenomena that occur under carefully controlled conditions, such as ideal concentrations of growth factors that have predictable effects on selected cell lines, and practical situations that involve the complex of mammalian systems and a plethora of different growth factors (both stimulatory and inhibitory) in environments complicated by infection, tissue necrosis, and external extremes. Although several angiogenic growth factors have been identified, controlling their activity in vivo remains elusive, probably because we lack understanding of the extracellular milieu of growth factors in vivo. Although the sources of growth factors have been identified (e.g., endothelial cells, macrophages, fibroblasts), the mechanisms that stimulate their controlled release and the three-dimensional ultrastructure in which they naturally reside are not well understood. It should therefore not be surprising that growth factors attached to synthetic polymers like polylactic acid and Marlex mesh are not particularly effective. Similarly, bioartificial membranes comprised of selected molecules, such as hyaluronic acid or purified Type I collagen laced with a variety of growth factors, usually fail to produce the desired effect in clinical situations.

Many synthetic and natural materials have been investigated for treating wounds in both military and civilian applications (e.g., Germain and Auger, 1995). Among these are wide classes like biodegradable polymers and biomodified materials that slowly release growth factors, blood-clotting agents, angiogenic inductors, or other potentially beneficial molecules. Despite the multitude of approaches and the investment of significant private and public resources, wound healing by material apposition is still far from being achieved.

Two types of possible solutions may be envisioned; both require significant advances in materials science and technology. The first solution is better materials, specifically materials that can arrest blood loss, impede infection, counteract shock, and foster biological regeneration. Multiplicity of function would presumably require composite materials, comprising a single or multiple matrix, with interdispersed biological molecules that are released over desired periods of time. The second solution, favored in the

NRC (2001) report, is to use biological self-replicating systems (e.g., Gentzkow et al., 1996). These systems integrate living cells into synthetic scaffolds for in vitro generation of the desired product (e.g., skin replacement) or even implantation of regenerating bioreactors at the wound site. Major breakthroughs in biomaterials science and engineering are needed to produce scaffolds that allow for such cell/material hybrid bioreactor to perform their desired functions in vivo.

Human Performance Enhancement

Though any material or technology can be viewed as a human performance enhancement, this review will concentrate on direct modifications to the human body. An example is the use of materials to strengthen the living skeleton. Applying new materials science to the management of human disease would certainly help DoD to maintain the health and readiness for duty of personnel. However, in this section we are considering the role of materials in increasing human performance substantially beyond the optimal native condition of the human body. This panel believes that many operations would have dramatically improved chances for success if nominal body performance is enhanced.

Examination of current enhancements/aids to the human body is informative. The enhancements shown in Table 7-2 are all heavily dependent on materials science. Materials science will continue to be critical in further advances, which will also be important for disease management.

While a body systems approach might be considered, the panel chose to examine possibilities by functional capability. An advantage of this approach is that optimizing overall performance has to be considered even while trade-offs become apparent. For example, if we want a soldier to be able to run faster, we will have to consider strength to weight in skeletal structure, improved muscle-tendon-bone attachments, improved flexibility and lubrication of moving body parts, changed energy use and waste production, the physical size of the limbs, and the effects on lifespan of body parts and the individual. Table 7-3 considers a reasonable number of functions and points out areas of needed research. Much enabling technology for advances in human performance comes from materials specialists. Such advances include materials for structure and function to create artificial tissues, the ability to manipulate cells and add biochemical groups to them, and advances in microfluidics to allow precision release of physiologic and pharmacologic molecules. Embedded sensor, analog, and logic circuitry (smart materials) will be needed.

TABLE 7-2 Current Human Enhancements and the Materials Enhancements They Depend On

Human Enhancement Technology	Materials Enhancement
Contact and intraocular lenses	Transparency, fluid biocompatibility
Hearing aids	Ear mold softness, microcircuits, batteries
Cochlear prostheses	Electrode compatibility
Tooth implants	Hardness, durability, anchoring
Larynx implants	Elasticity
Silicone implants	Biocompatibility, natural feeling
Joint and bone replacements	Hardness, durability, bone compatibility
Artificial blood	Membrane permeability, ductility
Vascular pumps, implants, and stents	Vascular lining, flexibility
Renal dialysis	Membrane permeability

TABLE 7-3 Human Body Functions That Could Potentially Be Enhanced and the Materials Advances Required

Body Function	Materials Advances Required
Locomotion/muscular activity	Energy sources for muscles, electronic implants for muscle tone, increased structural integrity for tendons and bone: microfluidics, micromaterial engineering
Nervous system control signals	Understanding signal transmission and materials improvements for nerves: tissue engineering
Energy management	Energy uptake, enhanced long-term delivery to tissue, improved waste removal: molecular coatings, enhanced excretion
Oxygen management	Improved gas exchange for uptake and excretion, artificial blood: surfactant development, tissue surrogates
Physical integrity	Skin, eye, and mucous membrane protection: tissue integrity, protection from radiation
Thermoregulation	Heat retention and rejection, energy recycling: smart materials that recognize the external environment
Cognition/mental states	Energy and toxin management for the central nervous system: drug delivery systems

To take advantage of the advances in materials technology in the area of human performance, there needs to be an interactive review of the issues and an awareness of the possibilities. Human performance specialists need to provide materials specialists with information on limiting factors. This may require new studies to better define the limitations. For example, physiologists should be able to determine if supranormal walking endurance will be limited by delivery of oxygen and energy or buildup of lactic acid. In a complementary fashion, advanced materials need to be described to physiologists so that they can contemplate how to take advantage of the advances. Human performance enhancement as a new area of work particularly needs an informed multidisciplinary approach. A specific forum must be created to take the specialists beyond their academic disciplines or even the disease mitigation mindset (i.e., research in medicine) and give them the information to be creative.

RESEARCH AND DEVELOPMENT PRIORITIES

Biology can affect every area of DoD needs of the next several decades. In some areas, i.e., improved battlefield medicine and biological warfare agent identification or interdiction, solutions may be found in specific biological molecules, cells, or systems; in other cases, i.e., smart materials or lightweight structural materials, biology may point the way to improved strategies for material design and synthesis. Where biological molecules or cells are the active component of a device, the challenge to the materials community is not only incorporation of the sensing entity but, perhaps more important, the preservation of biological function in a nonbiological environment. The common theme for all these technologies, research areas, and applications is using biological paradigms to solve problems of materials design, materials synthesis, and system assembly.

The panel selected the following research and research management priorities as critical for realizing the opportunities for bioinspired and bioderived materials research for meeting future defense needs.

Improving Fundamental Understanding of the Relationships Between Biological Structure, Properties, and Evolution and Materials Design and Synthesis

Recent studies of biological systems have clearly shown that large numbers of molecules, structures, and systems in living organisms possess attractive materials properties that are beyond the reach of current nonbiological synthetic approaches. Examples are as follows:

- The natural fabrication processes of hierarchical, systems-oriented biological structures, spanning the size range from nanometers to the macroscopic, lead to impressive and highly desirable performance.
- The combination of mechanical properties displayed by bone and the instantaneous, highly selective, single-molecule sensitivity of biosensing systems on cell membranes are two examples of the attractive properties and performance of hierarchical, biological systems.
- Many molecules, structures, systems, and natural fabrication processes have the potential to serve as the basis for materials with enhanced properties for defense applications, either directly adapted or as a pattern for nonbiological mimics.

Increasing Communication of DoD Material Needs to Biological and Physical Scientists

Progress at the interface between biological and physical scientists is hampered by educational differences and a general lack of communication, especially the lack of communication of DoD materials needs to the biological community. Biology offers a rich source of strategies for solving material problems; conversely, the materials science paradigm offers a systematic methodology for identifying biologically relevant materials. Increasing communication across disciplinary boundaries is likely to produce dramatic benefits to both communities.

Basic Research into Biological Molecules, Structures, Systems, and Processes to Lay the Groundwork for Their Use, or Their Use as Models, in Serving the Materials Needs of DoD

Fundamental understanding of the relationships between biological structure, properties, and materials synthesis is required if DoD is to take advantage of any new bioinspired or bioderived material. Biological toughening of materials, as exemplified by nacre, tendon, and bone, offers useful models for the next generation of armor/damage-resistant materials. Biological control of primary structure and the resultant enhanced performance, as exemplified by the toughness of dragline silk, is a paradigm for materials by design.

Identification and Development of Biocompatible Materials to Enable in Vivo Implantable Devices

In vivo detection strategies to identify toxins and pathogens, including masked agents, may enable the detection of a single agent molecule. New materials for implantable, multifunctional, and tissue-interactive devices are required: Emphasis should be placed on materials for the microfluidic movement of biologically active molecules to specific tissues and on materials to interface with electrically active tissues.

Development of Packaging Technologies to Preserve the Biological Function of Biologically Enabled Devices

Preservation of the biological function of biological molecules is a key driver for the next generation of biologically enabled devices. Because current strategies are inadequate, this area must have high priority.

REFERENCES

Aizenberg, J., A.J. Black, and G.M. Whitesides. 1998. Controlling local disorder in self-assembled monolayers by patterning the topography of their metallic supports. Nature 394(6696):868-871.

Albert, J.T., O.C. Friedrich, H.E. Dechant, and F.G. Barth. 2001. Arthropod touch reception: Spider hair as rapid as touch detectors. J. Comp. Phys. A—Sens. Neural and Beh. Phys. 187(4):303-312.

Barth, F.G. 2002. A Spider's World: Senses and Behavior. Berlin: Springer-Verlag.

Bashir, R., R. Gomez, A. Sarikaya, M.R. Ladisch, J. Sturgis, and J.P. Robinson. 2001. Adsorption of avidin on microfabricated surfaces for protein biochip applications. Biotechnology and Bioengineering 73(4):324-328.

Chen, T., S.C. Barton, G. Binyamin, Z. Gao, Y. Zhang, H. Kim, and A. Heller. 2001. A miniature biofuel cell. J. Am. Chem. Soc. 123(35):8630-8631.

Christel, L.A., K. Petersen, W.A. McMillian, and M.A. Northrup. 1998. Rapid, automated nucleic acid probe assays using silicon microstructures for nucleic acid concentration. Journal of Biomedical Engineering 121(1):22-27.

Desai, T.A., D.J. Hansford, L. Kulinsky, A.H. Nashat, G. Rasi, J. Tu, Y. Wang, M. Zhang, and M. Ferrari. 1999. Nanopore technology for biomedical applications. Biomedical Microdevices 2(1):11-40.

Gentzkow, G.D., S.D. Iwasaki, K.S. Hershon, M. Mengel, J.J. Prendergast, J.J. Ricotta, D.P. Steed, and S. Lipkin. 1996. Use of dermgraft, a cultured human dermis, to treat diabetic root ulcers. Diabetes Care 19(4):350-354.

Germain, L., and F.A. Auger. 1995. Tissue engineered biomaterials: Biological and mechanical characteristics. Pp. 699-723 in Encyclopedic Handbook of Biomaterials and Bioengineering, Part B: Applications, Vol. 1, D.L. Wise, ed. New York: Marcel Dekker.

Harrison, D.J., A. Manz, Z. Fan, H. Ludi, and H.M. Widmer. 1992. Capillary electrophoresis and sample injection systems integrated on a planar glass chip. Anal. Chem. 64(17):1926-1932.

Heringdorf, F.J.M.Z., M.C. Reuter, and R.M. Tromp. 2001. Growth dynamics of pentacene thin films. Nature 412(6846):517-520.

Kaplan, D.L. 2002. Spiderless spider webs. Nature Biotechnology 20:239-240.

Koehl, M.A.R., J.R. Koseff, J.P. Crimaldi, M.G. McCay, T. Cooper, M.B. Wiley, and P.A. Moore. 2001. Lobster sniffing: Antennule design and hydrodynamic filtering of information in an odor plume. Science 294(5548):1948-1951.

Lakes, R. 1993. Materials with structural hierarchy. Nature 361(6412):511-515.

Lazaris, A., S. Arcidiacono, Y. Huang, J.F. Zhou, F. Duguay, N. Chretien, E.A. Welsh, J.W. Soares, and C.N. Karatzas. 2002. Spider silk fibers spun from soluble recombinant silk produced in mammalian cells. Science 295(5554):472-476.

McKenzie, I., R. Jones, I. Marshall, and S. Galea. 2000. Optical fibre sensors for health monitoring of bonded repair systems. Compos. Struct. 50(4):405-416.

National Research Council. 2001. Opportunities in Biotechnology for Future Army Applications. Washington, DC: National Academy Press. Available online at <www.nap.edu/catalog/10142.html>. Accessed January 15, 2002.

Siegel, R.W. 1993. Exploring mesoscopia—The bold new-world of nanostructures. Physics Today 46(10):64-68.

Wainwright, S.A., W.D. Biggs, J.D. Currey, and J.M. Gosline. 1982. Mechanical Design in Organisms. Princeton, NJ: Princeton University Press.

Wise, D.L. 2000. Biomaterials and Bioengineering Handbook, D.L. Wise, ed. New York: Marcel Dekker.

CHAPTER EIGHT

Integration of Research Opportunities

INTRODUCTION

The Committee on Materials Research for Defense After Next examined a broad range of materials research areas, including such diverse areas as bioderived materials for wound healing, high-temperature structural materials for advanced jet engines, and materials for advanced explosives and propellants. Despite this diversity, the recommendations resulting from this effort can be summarized into four principal themes:

- Design of materials, devices, and systems assisted by computation and phenomenological models of materials and materials behavior;
- Convergence, combination, and integration of biological, organic, semiconductor/photonic, and structural materials;
- Discovery and characterization of new materials with unique or substantially improved (50 percent) properties; and
- New strategies for synthesis, manufacture, inspection, and maintenance of materials and systems.

These themes are explored in detail in the sections that follow.

DESIGN OF MATERIALS, DEVICES, AND SYSTEMS ASSISTED BY COMPUTATION AND PHENOMENOLOGICAL MODELS OF MATERIALS AND MATERIALS BEHAVIOR

Throughout history, humans have continually sought to improve their existence by finding novel approaches to the discovery and application of useful materials. Early man refined processes for melting and shaping metals and their alloys through exhaustive trial-and-error approaches extending over millennia, giving rise to the Bronze and Iron Ages. Such trial-and-error processes have continued into modern times. The protracted search for an acceptable electric light bulb material culminated in the tungsten filament and contributed a new phrase to the English language: "Edisonian approach." The pace of materials discovery continues to quicken. Today's materials scientists have at their disposal increasingly powerful tools. Foremost among them is the breathtaking improvement in computational power, leading to the ability to predict certain structures from first principles. Indeed, in some cases, such as development of insensitive energetic materials, computational approaches are leading experiments in new materials synthesis and are reducing the time required to discover and apply new materials. Coupled with the use of phenomenological models to help identify and predict the characteristics and behavior of potentially revolutionary materials, advanced computational approaches offer a rapid and powerful means for discovery.

RECOMMENDATION 2. THE DEPARTMENT OF DEFENSE SHOULD MAKE RESEARCH INVESTMENTS IN THE DESIGN OF MATERIALS, DEVICES, AND SYSTEMS ASSISTED BY COMPUTATION AND PHENOMENOLOGICAL MODELS OF MATERIALS AND MATERIALS BEHAVIOR.

The potential benefits from such investments for DOD could be staggering; a few of the opportunities and applications discussed in this report are outlined below:

- *In advanced structural composites* strength, toughness, durability, and damage tolerance are highly dependent on the particle-matrix or fiber-matrix interface. Reliable tailoring of composite characteristics to specific applications requires detailed understanding of these interfaces, which will come only after there are models for interface structures, properties, and thermodynamic stability. Success would lead to reduced need to overdesign the structural components of military systems, with a concomitant reduction in weight and energy consumption.
- *Advanced photonic devices* such as optical analog/digital (A/D) converters promise processing speeds significantly faster than their electronic counterparts, enabling advanced optical networking and communi-

cations. This capability will require materials with 500 times higher electro-optical coefficients than current materials like $LiNbO_3$. Such properties can probably be achieved only in artificially structured materials that are unlikely to be found "by accident." Enhanced theory and modeling will be essential in guiding experimental efforts if these promising compositions and structures are to be identified and applied.

- *Protection of eyes and sensors against battlefield lasers* is important across the DoD spectrum. This problem has been exacerbated by the advent of tunable lasers, making multiwavelength protection a necessity. Use of a priori quantum chemical models combined with phenomenological models in the design of advanced materials has led to increased third-order nonlinear susceptibilities and two-photon absorption cross sections. Enhanced modeling would assist in the discovery of new families of optical materials for protecting eyes and sensors from hostile or accidental laser beam impingement over wide ranges of frequency and flux intensities.
- *Lightweight mobile power* is particularly important for the Army and the Marine Corps. Advanced computational techniques combined with phenomenological models should assist in identifying macromolecules for advanced thermoelectric materials that possess higher figures of merit. DoD could then take advantage of waste heat to generate mobile power and could also miniaturize electronic and EO components through use of positive thermal management protocols.
- *Cost of ownership of military vehicles* over their lifetime is very important to DoD. Prediction and extension of component life would remarkably reduce this cost but will be possible only as computational models are created for the aggregate properties of reliability, durability, and affordability.
- *There is a veritable cornucopia of inorganic materials* that are promising for future DoD applications, ranging from lean alloys for structural use to wide bandgap semiconductors for high-power electronic applications. Determining which are most appropriate to DoD needs depends on understanding the mechanisms by which trace elements and individual defects affect key materials properties. Such understanding will come only from the use of appropriate computational models.
- *The computational power* required to support all these needs was unimagined a generation ago, particularly the massively parallel architectures of today's most powerful computers. The concept of massive parallelism also applies on the experimental front. For example, combinatorial synthesis (the automated parallel examination of many potential solutions

with criteria for identification of best options built in) is inspired by the deeply parallel discovery processes found in living systems. Examples of applications of combinatorial synthesis range from drug discovery to identification of high-temperature superconductors. Real-time application of combinatorial synthesis could be the basis for identifying agents to counter chemical or biological agents.

CONVERGENCE, COMBINATION, AND INTEGRATION OF BIOLOGICAL, ORGANIC, SEMICONDUCTOR/PHOTONIC, AND STRUCTURAL MATERIALS

Convergence, combination, and integration are major themes for addressing future defense needs. History has shown that major advances usually occur at intersections or points of convergence between disparate areas. This is true in both broad fields of endeavor (e.g., chemistry, physics, biology) or more specific areas (e.g., lithography, microsystems).

RECOMMENDATION 3. THE DEPARTMENT OF DEFENSE SHOULD MAKE RESEARCH INVESTMENTS THAT PROMOTE CONVERGENCE, COMBINATION, AND INTEGRATION OF BIOLOGICAL, ORGANIC, SEMICONDUCTOR, PHOTONIC, AND STRUCTURAL MATERIALS.

Examples of the most significant opportunities for investments that support the future needs of the DoD are provided below.

Convergence

The convergence of biology with traditional materials science with the application of biological principles to the design of structural materials and microsystems is an area particularly ripe for revolutionary advances in materials for DoD. The concept of bioinspired processing applies lessons from biology to creation of synthetic materials. This approach is ideally suited for the design and fabrication of nanostructured organic and organic/inorganic composites. Materials like bone, teeth, and shells are simultaneously hard, strong, and tough and have unique hierarchical structural motifs originating at the nanometer scale. Mimicking such designs should lead to very strong, tough materials usable in, e.g., lightweight armor for both warfighters and vehicles. It could also be used for mechanical system components.

Examining the lessons of biology allows us to envision microsystems based on entirely different architectures than are used today. While today's microsystems use components like gears whose features are essentially scaled down from macroscopic dimensions, future microsystems may be built from components that are either inspired by the ways that biological systems work or that exploit actual biological materials. DoD should actively seek first to identify and then to exploit such points of convergence between materials science and biology.

Combination

Many major advances over the history of materials resulted from synergistic combinations of materials that yield property levels that cannot be obtained in a single material. Once it was recognized that the presence of other atoms could lead to improved performance, the use of metals in ancient times quickly gave rise to intentional alloying for structural applications. Similarly, bricks made from mud and straw proved more durable than mud alone. In the latter half of the 20th century, advanced composites incorporating polymers, metals, and ceramics reinforced with other materials in fiber, whisker, platelet, and other geometries gave the structural designer new possibilities. Because many material combinations have already been explored, the probability of obtaining substantial improvements in properties through empirical combination of different materials and geometries has been decreasing. However, a resurgence is under way due to exploitation of a new variable: size scale. In particular, there is the potential for major improvements in performance from intimate and controlled mixing of materials at the nanoscale. This area provides fertile ground for significant advancements in many material properties of importance to DoD.

While the practice of combining materials, each adding its own unique characteristics and functionality, has been widely explored in structural materials, exploration of combinations for other applications is only beginning. An example is the use of high-yield photon absorbers with materials that are highly efficient charge carriers, vastly improving solar-energy harvesting compared with a single material. Material combinations in nontraditional areas provide a compelling strategy in the search for next-generation functional materials.

Combining organic and inorganic materials into composites to provide new ionically conducting pathways in polymer battery systems, modifying the current mechanism responsible for ionic mobility, offers DoD many potential benefits. Nanostructured clays in polyethylene oxide (PEO) and nanoparticulate oxides in PEO have both yielded improved conductivity and transference number. Composites can alter the mechanism of ionic transport, suggesting the possibility of high conductivity, with ionic transport number of the "working ion" (e.g., Li^+) near unity.

Integration

Integration is a third important subtheme. Whereas combination connotes the union of two or more materials to achieve a characteristic or property level not possible using a single material, integration is the purposeful, seamless union of separate functions in a material, device, or structure that contributes to true multifunctionality. Function can be integrated at different size scales; multiple functions can be integrated in several materials or in a single material.

At the macroscale, a traditional structural composite contains more than one material, but it is a combination that has only one primary function, i.e., to carry loads and moments. A more advanced, integrated composite might, in addition to providing structural capability, be able to absorb radar wavelengths or to effect local color change where damage has occurred ("bruising") to alert maintenance personnel.

The integration of multiple functions at the macroscale, the microscale, and the nanoscale will likely be characteristic of the DoD systems of 2020. As an example, an almost limitless number of organic composites can be made by combining matrix resins that have different bonding capabilities (e.g., trifunctionality, tetrafunctionality) and functional groups that have multifunctional complementary phases. This may lead to structural materials that can also absorb radar energy, conduct electricity, serve as large area sensors, or perform other useful functions.

Another example is monolithically integrated microsystems on a warfighter's clothing or gear that would sense exposure to a chemical or biological agent, diagnose the type and degree of exposure, and then fabricate and administer an antidote. This might require integration of both materials and functions, e.g., semiconductors (for logic), optical materials (for sensing), organic or biological materials (as molecular recognition sites for chemical or biological agents or for another analytical technique), and organic or biological materials (for fabricating the antidote).

A limiting factor in the use of biological molecules or cells as the active component of sensing or diagnostic devices is preservation of the biological function and stabilization of biological activity in a non-biological environment for extended periods of time. Packaging strategies will be needed to assure continued activity of the biological components of such devices.

Integration of function at the nanoscale in organic materials for computational devices has been of interest for some time. Most, if not all, of the individual molecular-scale building blocks that are required to create electronic circuits have recently been demonstrated, although reproducibility has been problematic. The next step is to create economically viable and highly reliable electronic circuits that are equivalent or superior to silicon technology but operate on a molecular scale. A number of concepts and designs on how to build basic molecular-scale electronic adders or logic circuits have been put forward. To make these a reality, advanced computational design and then synthesis of durable molecular organic, polymer, and hybrid building blocks will be necessary, coupled with advanced fabrication, connection, and assembly technologies.

A key aspect of materials science that must be considered in both combination and integration of multiple materials is the importance of preserving material performance at interfaces. Nowhere is this more clearly illustrated than in the advanced electronic and photonic devices that could revolutionize computer, optical, and microsystem architectures. Successful realization of quantum electronic devices will live or die with our ability to control atomic-level composition and atomic placement in immediate proximity to interfaces, and to control overall device dimensions with the same degree of accuracy. The semiconductor industry already accomplishes this routinely in one dimension, but future devices will require the same degree of control in the other two dimensions as well. Devices that operate at wavelengths of interest and use photonic bandgap engineering will likely not be as sensitive to atomic-level accuracy, but the materials that need to be integrated will probably be more incompatible chemically. This challenge is likely to be equally as great as the challenges of quantum electronic devices.

Processing and characterization are integral to success in any merging of materials or functions. For example, 21st-century devices may require the union of rigid inorganic materials that have hitherto required high processing temperatures with synthetic or even biological macromolecular structures that have drastically different material characteristics. Processing such hybrid structures presents a tremendous challenge. The interfaces between materials must be clearly understood because functionality is often derived from the materials properties very close to an interface. As device and system dimensions shrink, interfaces become even more dominant because an increasing proportion of atoms or molecules find themselves at one interface or another. Advances in methods to characterize interfaces, both during processing and after deployment, coupled with predictive models that enable interface property optimization, will increase in importance. Solving this type of problem will necessitate addressing all four of the research opportunity themes identified by the committee and discussed in this chapter.

DISCOVERY AND CHARACTERIZATION OF NEW MATERIALS WITH UNIQUE OR SUBSTANTIALLY IMPROVED PROPERTIES

DoD systems of 2020 would benefit significantly from the discovery, production, and application of materials that have either unique properties or properties that considerably exceed those of today's materials. DoD weapon systems and platforms must be lethal and sustainable, and they must enhance the survivability of the user. These requirements continue to drive new materials discovery.

Reducing volume, mass, or both while enhancing functionality are key system drivers. Munitions must be more compact; power and energy sources must have higher volumetric and mass densities; armor must be lighter while providing equivalent or enhanced protection; platform structure should be lighter but remain strong so that payload can be increased; and electronic and optical communication systems must be smaller and lighter but have higher capability and bandwidth. All these are tangible system requirements that depend on the discovery and evolution of substantially improved new materials in subsystems, device components, and subcomponents.

RECOMMENDATION 4. THE DEPARTMENT OF DEFENSE SHOULD MAKE RESEARCH INVESTMENTS THAT PROMOTE DISCOVERY AND CHARACTERIZATION OF NEW MATERIALS WITH UNIQUE OR SUBSTANTIALLY IMPROVED PROPERTIES (BY 50 PERCENT OVER CURRENT PROPERTIES).

Examples of the most significant opportunities for such investments that support the future needs of the DoD are provided below.

Design and fabrication of tunable materials with low-loss dielectric and piezoelectric properties would enable low-loss phase shifters that could operate at the high frequencies required for IR countermeasure systems. Improved organic photovoltaic materials that have higher efficiencies could permit localized power harvesting and generation, providing increased battlefield mobility at relatively low cost. High-quantum-efficiency electroluminescent materials, assuming 50 percent improvement over today's levels, will permit organic light-emitting materials to compete effectively as efficient large-area solid-state light sources and in flat-panel displays. This would have far-reaching benefits for DoD in terms of savings in weight and cost and increased battlefield ruggedness.

Energetic materials also offer the potential for substantially improved (greater than 50 percent) properties. For example, polyatomic nitrogen compounds could yield propellants with greatly enhanced specific impulse, and could also be used in enhanced explosives. Novel salts of polyatomic nitrogen species might be able to provide a 5-6 times mass reduction in energetic material for the same payload mass in missiles, allowing for huge performance increases.

While highly nonlinear effects in traditional bulk materials are unlikely, nonlinear optical materials offer the potential for substantially improved optical limiting capabilities. Strong confinement of electrons in organic materials or quantum dots may lead to enhanced nonlinear optical coefficients. Nanostructured material approaches could provide wholly new routes to solving the considerable problem of protecting eyes and sensors from laser radiation at a variety of frequencies. Computational and combinatorial approaches may accelerate the pace of developing such materials.

Novel materials with unique properties could also lead to tunable IR detectors capable of operating across the 1- to 15-micron region. Carbon nanotubes, whose semiconducting bandgaps can range from less than 100 meV to several hundred meV, offer a new approach to tuned materials; the ability to control bandgap by controlling tube diameter and chirality of nanotubes in a well-aligned film offers tantalizing possibilities.

Binding site-receptor interactions in biology offer a strategy for the discovery of useful active agents for interdicting unknown toxins and pathogens while identifying known pathogens. For example, antibodies active against an unknown agent may be identified with recombinant DNA techniques while an antibody known to bind to a specific pathogen may be used as the active site of a sensor for that pathogen.

Techniques for remote detection of biological agents and pollutants continue to interest DoD. Bacteria, viruses, and many bacterial toxins, as well as all known nutrient media for biological agents, have strong UV fluorescence spectra that can serve as indicators for remote biodetection. Efficient UV laser and detector media are largely materials-limited. GaN-based wide-bandgap semiconductors are promising for blue and UV lasers, but because a fundamental understanding of these materials is lacking, how they respond to DoD needs is not established.

Membrane research will continue to be a fertile area for DoD over the next two decades. Membranes are ubiquitous in numerous applications, among them chemical-biological defense, water purification, and energy technology (e.g., fuel cells). The fundamental science of modeling, synthesis, and processing of membrane materials could lead to smart membrane materials that could report on the local environment and change properties as needed, as well as repair their own defects. Other advances could enable energy-efficient water cleaning and ion-transporting membranes for high-performance power generation.

NEW STRATEGIES FOR PROCESSING, MANUFACTURE, INSPECTION, AND MAINTENANCE OF MATERIALS AND SYSTEMS

The themes already discussed describe an approach to the discovery of novel materials, combinations of materials, and integrated functionalities. Discovery of their structures and potential properties is an important step toward realizing their potential for new and far-ranging DoD capabilities. However, this is only the starting point for introducing materials into systems. Materials must be processed and characterized successfully in the laboratory on a small scale, and then practical and robust methods for scale-up to sufficient quantities must be applied. Techniques must be available for fabricating and assembling the materials into prescribed geometries, and for quality assurance to make certain they meet requirements. Finally, there must be techniques for inspecting them in the field to identify problems, repair them, and assure the adequacy of repair. The panel reports have suggested a number of areas that could help assure the successful transition of new materials into defense applications.

RECOMMENDATION 5. THE DEPARTMENT OF DEFENSE SHOULD MAKE INVESTMENTS IN RESEARCH LEADING TO NEW STRATEGIES FOR THE PROCESSING, MANUFACTURE, INSPECTION, AND MAINTENANCE OF MATERIALS AND SYSTEMS.

Examples of the most significant opportunities for such investments to respond to future needs of DoD are provided below.

The capability for continuous characterization of materials while they are in service could provide significant benefits to DoD. Military systems require predictable high-level performance in a range of adverse environments. In-service characterization would allow for continuous monitoring of the health of systems and would provide essential information on how key properties change as a result of service-induced damage. One of the key enablers for this capability would be smaller, more sensitive NDI/NDE sensors that can be attached to or included in structural materials and the structures themselves when they are manufactured. Such sensors could allow for continuous monitoring and real-time recovery of information from multifunctional material systems for use in evaluating safety, need for replacement or maintenance, or remaining life. Reducing the large amounts of data collected by such sensors into much smaller amounts of useful information will be a challenge.

Repair, maintenance, and life extension of military systems is an important current and future issue for DoD. Materials that self-repair or heal themselves could pay major dividends, at least for localized damage if there were a practical means for incorporating self-repair mechanisms in them. The feasibility of restoring the strength of matrix materials using embedded chemicals has already been demonstrated. However, other approaches, if fully realized, might lead to a wide array of self-healing polymers, composites, and adhesives for structures. A critical aspect to be addressed is decreasing the time required for healing response so that it is rapid enough to be useful.

Purity will be increasingly important for multifunctional, organic, electro-optically active materials capable of high performance and reliability. This will require routine, affordable synthetic processing strategies that reduce chemical and morphological defects to levels that are currently at best barely achievable. The future of macromolecular materials in high-performance DoD applications depends on processing of high-purity, well-defined, reproducible structures that are affordable. Likewise, fielding new materials with extreme properties will require emphasizing, first, understanding their characteristics and, then, controlling them through fabrication. This emphasis will allow for production of components and subsystems for frequency-agile communication and radar applications that take advantage of the field-variable properties of ferroelectrics, ferrites, and other new materials.

Similarly, accelerated DoD systems application of solar-energy-harvesting devices would benefit from low-cost manufacturing of materials and devices that provide efficient photon collection and rapid charge conduction, and that can be applied over large areas. As noted in Chapter 6, organic photovoltaics are one possible solution that also allows use of flexible substrates. If such approaches are to become widely applied in military systems, a cost-effective way to make photovoltaic cells in large quantities is essential.Today's laboratory-scale techniques for generating promising material configurations must give way to truly manufacturable approaches.

The promise of nanomaterials for a wide variety of DoD applications has been discussed throughout this report. Today, such materials are barely more than a laboratory curiosity. Usually, only excruciatingly small quantities of material can be produced and production is generally measured in hours or days. If nanomaterials are to become widely applied, it will be essential to address the challenge of scaling up their production to macroscopic quantities while retaining the performance properties of the small samples currently available. Some properties of nanomaterials also need to be better understood so that they can be engineered for specific applications.

Biomimetic and bioinspired manufacturing approaches have considerable potential for meeting future defense needs. Much current research, particularly in biotechnology, has been oriented toward pharmaceutical and medical products; biomanufacturing approaches can also apply to the nonmedical arena. For example, biocatalysis is an exceptionally efficient manufacturing process that can accelerate reactions by up to 13 orders of magnitude; it has exquisite specificity for starting materials and products (no byproducts) at room temperature and atmospheric pressure. The activity of enzymes can be controlled over several orders of magnitude by binding specific effector molecules. Selective activation of enzymes could effect specific chemical conversions in materials for munitions, therapeutic agents, odorants, and other uses. Indeed, harnessing such approaches would be revolutionary.

Materials are rarely applied in isolation; usually, a system of materials must work together seamlessly. As an example, a fuel cell contains such elements as electrodes, membrane, electrolyte, and catalyst. Optimal application of new materials approaches (e.g., nanostructured electrodes, polymeric membranes) to this complex electrochemical system will require methods for characterizing the structure and stability of the individual materials as well as their interaction within the system as a function of such factors as time, temperature, and local chemical environment. Accomplishing this in the laboratory is only the first step. Cost-effective, reliable synthesis and manufacturing of the individual components must be coupled with appropriate assembly and quality assurance techniques, followed by testing in military environments.

There are many steps between laboratory demonstration of a single material and successful introduction of a system of materials into a military application where problems can arise. The committee strongly recommends that a life-cycle view be taken as soon as possible in the materials design process so that questions about application-relevant characteristics can be formulated and answered early. This will help systems designers to avoid individual materials or materials combinations that are unsuitable and that could lead to major difficulties in a system or subsystem. Short term, this will clearly require spending time and other resources to resolve uncertainties about materials options; long term, this approach should reduce the life-cycle cost of maintenance and repair.

CONCLUSIONS

Materials science and engineering is entering a renaissance due to the advent of enhanced tools for discovering, processing, structuring, combining, and characterizing matter in wholly new ways and at ever-decreasing scales. This has profound implications for DoD.

Today's DoD requires technology that yields high performance and is reliable under extremely adverse conditions. Tomorrow's DoD will rely on tremendous advances in technology. Intelligence will be gathered using large numbers of unmanned autonomous vehicles and sensors that are fully networked, allowing the battlespace to be monitored from different vantage points—undersea, on the ground, in the air, and from low Earth orbit. Battles will be fought using more lethal weapon systems that have tailorable effects and will require fewer but more highly skilled and better-equipped soldiers, sailors, and aviators. Transformation of the way the United States chooses to fight will eliminate heavy, slower systems and place a premium on speed and tactics, as well as communication. This will necessitate major improvements to weapon platforms that increase mobility and enhance survivability. Except for software, most of these advances for 2020 will be enabled by improvements in materials, processing methods, and associated design techniques.

To achieve these capabilities, DoD should emphasize the four overriding themes described in this chapter. First, computational and phenomenological models of materials and their behavior are essential to the design of both novel materials themselves and the devices and systems incorporating them. Second, convergence, combination, and integration are essential. DoD should seek out the intersections of major bodies of knowledge, for it is here that major advances disrupting the status quo are often found. Of particular interest is the intersection of biology and materials science. Combination of materials in many forms, taking advantage of new approaches to specifically tailor structure at different length scales, should be pursued. Integration of biological, organic, semiconductor, and structural materials should be examined as a way to obtain true multifunctionality at both the macroscale and the microscale. Third, DoD should seek to discover and characterize materials that have either unique or substantially improved properties. The numerous possibilities identified in the reports of the panels span the gamut of DoD applications and involve a broad range of materials. Foremost among these is the exploitation of our increasing ability to control the structure of materials at the nanoscale. Fourth, new strategies for the synthesis, manufacture, inspection, and maintenance of materials and systems are essential.

Materials science and engineering may be entering a renaissance, but its potential for the DoD will be realized only if exciting small-scale advances in the laboratory can be translated into practical large-scale production processes and systems. Traditional stovepipe approaches to management of materials and processes within the materiel life cycle will result in delayed or missed opportunities for transition into fielded military hardware. Overcoming these historical barriers will require a sustained, concerted effort by DoD and the industrial managers responsible for advancing materials throughout the life cycle.

This shift in approach will undoubtedly require a painful cultural change. However, without it the pace of transition to new materials in DoD platforms and weapons systems will remain unacceptably slow, representing lost opportunities to make rapid, dramatic, and far-reaching changes to our military systems that meet constantly evolving and increasingly more sophisticated threats to the well-being of the nation.

Appendixes

APPENDIX A

Meeting Speakers

DECEMBER 6–7, 1999
DEFENSE SCIENCE AND TECHNOLOGY RELIANCE SUBAREA FOR MATERIALS AND PROCESSES WORKSHOP
ANNAPOLIS, MARYLAND

December 6, 1999

Overview of M/P Reliance Status	Lewis Sloter, ODDR&E
Army M&P Program Update	Dennis Viechnicki, ARL
Navy M&P Program Update	Robert Pohanka, ONR

December 7, 1999

Air Force M&P Program Update	Robert Rapson, AFRL
DARPA M&P Program Update	Steven Wax, DSO
BMDO M&P Program Update	James Shoemaker, BMDO
History of DoD Materials	Jerome Persh, IDA

Technical Panel (TP) Breakout Sessions with NMAB committee members

TP 1.0—Platform Structural Materials	Scott Tiebert, AFRL
TP 2.0—Power and Propulsion Materials	Kathy Stevens, AFRL
TP 3.0—Armor/AntiArmor Materials	Robert Dowding, ARL
TP 4.0—Electronic Materials	Bill Woody, AFRL
TP 5.0—Laser-Hardened Materials	George Mueller, NRL
TP 6.0 (A)—Operational Support Materials	Gume Rodriguez, ARL
TP 6.0 (B)—Nondestructive Evaluation	Ignatio Perez, NRL
TP 7.0—Signature Control Materials	Don Woodbury, ARL

FEBRUARY 15–16, 2000
NATIONAL RESEARCH COUNCIL
WASHINGTON, D.C.

February 15, 2000

Overview of DoD Vision and System Needs	Andrew Marshall, OSD
Overview of Army Vision and System Needs	Michael Andrews, U.S. Army
Overview of Military Transformation and Future Warfare	Michael Vickers, Center for Strategic and Budgetary Assessments
Overview of Air Force Vision and System Needs	Lawrence Delaney, U.S. Air Force
Overview of Navy Vision and System Needs	Ronald De Marco, ONR
Overview of Marines Vision and System Needs	Gen. Alan Gray, retired (U.S. Marine Corps)
Revolution in Military Affairs After Next	Lonnie Henley, Defense Intelligence Agency
Brainstorm Session on Systems Needs	Millard Firebaugh, Facilitator

February 16, 2000

Overview of ARO Materials Programs	Andrew Crowson, ARO
Overview of AFOSR Materials Programs	Lyle Schwartz, AFOSR
Air Force Chief Scientist Study (Overview of Air Force S&T Needs)	Kenneth Harwell, AFRL
Overview of NSF Materials Programs	Thomas Weber, NSF
Overview of DOE Materials Programs	Robert Gottschall, DOE
Brainstorm Session on Materials Areas	Julia Phillips, Facilitator

APRIL 18–20, 2000
ARNOLD AND MABEL BECKMAN CENTER
IRVINE, CALIFORNIA

April 18, 2000

Recap of Systems Needs Session from February Meeting	Millard Firebaugh
Summary of Navy Needs	Millard Firebaugh

Summary of Army Needs	John Gassner
Summary of Air Force Needs	George Peterson
Recap of Panels Structure from February Meeting	Julia Phillips
Structural Materials	Richard Tressler
Electronic/Photonic Materials	Meyya Meyyappan
BioMaterials	Michael Jaffe
Energy/Power Materials	John Gassner
Multifunctional (Other) Materials	Frank Karasz

April 19, 2000

Structural Materials Challenges for Long-Term Defense Needs	Anthony Evans, Princeton University
Materials Research at Bell Laboratories	Bertram Batlogg, Lucent Technologies
Biology as Our Mentor for Materials Research—Can We Learn from Its Example?	Mark Alper, University of California, Berkeley
Computational Materials Science to Meet Long-Term Needs	Stephen Foiles, Sandia National Laboratories
Carbon Nanotubes and Other Revolutionary Materials and Processes—Industry View	Fred Herman, Lockheed Martin
Emerging Organic Materials and Processes—Long-Term Challenges	Samuel Stupp, Northwestern University
Review of Enabling Crosscutting Areas for Materials Research	Millard Firebaugh, General Dynamics

April 20, 2000

Transition of New Materials to Development of Systems—Cost Perspective	Paul Kaminski, Technovation, Inc.
Power Source Materials and Technologies—Long-Term View	Daniel Doughty, Sandia National Laboratories

JUNE 28, 2000
J. ERIK JONSSON CENTER
WOODS HOLE, MASSACHUSETTS

June 28, 2000

Debriefing and Discussion—Out of the Box and Into the Future: A Dialogue Between Warfighters and Scientists on Far-Future Warfare (2025) Conference, June 26-27, 2000, Washington, D.C., organized by the Potomac Institute for Policy Studies	Sharon Yeung, National Research Council

MARCH 28-29, 2001
NATIONAL RESEARCH COUNCIL
WASHINGTON, D.C.

March 28, 2001

DoD Study Sponsors' Expectations and Feedback	Steven Wax, DARPA
Future DoD Systems and Impact of Materials Advances—Army Perspective	LTG Paul Kern, U.S. Army

March 29, 2001

Structural and Multifunctional Materials

Overview of Multifunctional Materials	Steve Wax, DARPA
Multifunctional Materials—Navy/DoD Perspective	Virginia DeGiorgi, NRL
Ceramic Structural Materials	Ron Kerans, AFRL
Metallic/Intermetallic Structural Materials	Dennis Dimiduk, AFRL
Polymer Structural Materials	Eric Wetzel, ARL
Metallic Structural Materials	Julie Christodoulou, ONR

Energy and Power Materials

Energy Conversion and Harvesting Materials	Bob Nowak, DARPA
High Power Microwaves and High Power Lasers	Tim Andreadis, NRL

Air Force Needs and Programs for Magnetic Materials	Richard Fingers, WPAFB
Explosives, Pyrotechnics, and Propellants	Brad Forch, ARL
Energy, Storage, and Thermal Sciences Branch Over	Dave Ryan, WPAFB

Electronic and Photonic Materials

Overview of Electronic (and Photonic) Materials	Valerie Browning, DARPA
Electronic Materials	Bill Mitchel, AFRL
Photonic Materials	Ken Hopkins, AFRL
Computational Materials Science	Dimitris Papaconstantopoulos, NRL
Computational Study of Strongly Correlated Materials	Steve Hellberg, NRL

Functional Organic and Hybrid Materials

Novel Organic and Organometallic Materials	Len Buckley, NRL
Functional Polymer Research	Matt Bratcher, ARL
Functional Organic Thin Films for Next-Generation Devices	Alberto Pique, NRL
Polymeric Materials	Doug Dudis, WPAFB

Bioderived and Bioinspired Materials

Biomimetics and Biosensor Research Within the Air Force Research Laboratory and Other DoD Agencies	Morley Stone, WPAFB
Biomaterials Prospects for the Navy and Marine Corps	Harold Bright, ONR
Materials and Devices: Biology Blurs the Distinction at the Nanoscale	Jeff Byers, NRL
Biologically Derived Structural and Functional Materials: Insight from Nature for Design Principles and Manufacturing Possibilities; and Nanobiotechnology for Chemical and Biological Defense	Robert Campbell, ARL

JUNE 26-27, 2001
J. ERIK JONSSON CENTER
WOODS HOLE, MASSACHUSETTS

June 26, 2001

Structural and Multifunctional Materials

Composite Materials in Systems	Frank Thompson, General Dynamics Land Systems
Metal Matrix Composites	Daniel Miracle, AFRL
Thermal Barrier Coatings	James A. Ruud, GE CR&D
Nondestructive Evaluations (NDE/NDI)	Michael Rooney, APL
Casting Geometrically Engineered Material Systems	Jon Priluck, JamCorp

Energy and Power Materials

Batteries and the Battery Industry by the Year 2025: An Industry Perspective on Outlook and Opportunities for DoD	Bob Staniewicz, SAFT America
Fuel Cells	Sossina Haile, California Institute of Technology

Functional Organic and Hybrid Materials

Organic Magnetics	Arthur Epstein, Ohio State University
Photonic Devices	Richard Ridgway, Battelle
Metal-Organic Catalysts	Bryan Coughlin, University of Massachusetts, Amherst

Bioderived and Bioinspired Materials

Review of the BAST Study on Opportunities in Biotechnology for Future Army Applications	Mauro Ferrari, Ohio State University
Foreign Developments in Biotechnology	Anna Pawul, Department of Defense
Applications of Structural Biopolymers in Defense and Industrial Systems	David Kaplan, Tufts University
Medical/Surgical Biomaterials and Devices	Steven Arnold, Johnson and Johnson

Human Performance Enhancements	Erik Viirre, University of California, San Diego

June 27, 2001

Structural and Multifunctional Materials	
Health Monitoring	Richard Claus, Virginia Polytechnic Institute and State University
Polymeric Materials	Richard Farris, University of Massachusetts, Amherst
Multiscale Materials	Priya Vashishta, Louisiana State University
Computational Materials Science	Michael Baskes, LANL
Energy and Power Materials	
Nanomaterials Science for Batteries and Other Power Sources	Rich Carlin, ONR
Energy and Power Sources for Future Army (and DoD) Applications	Bob Hamlen, Army CECOM
Fuel Cells	Al Tucker, ONR
Electronic and Photonic Materials	
Sensors	Mike Wazenski, Northrop Grumman
Ferroelectric/Piezoelectric Materials	Amar Bhalla, Pennsylvania State University
Microsystems	Don Cook, Sandia National Laboratories
Nanoelectronics	Charles Lieber, Harvard University
Communications/Wireless Systems	David Eaglesham, Lucent Technologies
Functional Organic and Hybrid Materials	
Organic Superconduction	Ananth Dodabalapur, Lucent Technologies
Actuators	Ray Baughman, Honeywell

OCTOBER 10-12, 2001
ARNOLD AND MABEL BECKMAN CENTER
IRVINE, CALIFORNIA

October 10, 2001

Structural and Multifunctional Materials

Nanostructured Materials	Julia Weertman, Northwestern University
Structural Amorphous Metals and Composites	William Johnson, California Institute of Technology
Ceramic Matrix Composites	Frank Zok, University of California, Santa Barbara
Manufacturing Ceramic Composites	Andy Szweda, COI Ceramics, Inc.
Structures/Joining	Bryan Dods, Boeing

Energy and Power Materials

Fuel Cells	Mark Williams, NETL
Micro Chemical and Thermal Systems (MicroCATS)	Robert Wegeng, PNNL
Capacitors	Chenniah Nanjudiah, Maxwell Laboratories
Fuels Processing	Rajeev Gautam, UOP
Future of Superconductors	Paul Grant, EPRI
Nanoenergetics	Rob Dye and H.E. Rogers, Technanogy

Electronic and Photonic Materials

Photonics/Laser Materials	David Hagan, University of Central Florida
Infrared/Radar Materials/Technologies	Brian Pierce, Raytheon
Thermoelectric Materials	Tim Hogan, Michigan State University
High-Temperature, High-Power Electronic Materials	Umesh Misra, University of California, Santa Barbara

Functional Organic and Hybrid Materials

Optically Active Polymers	Luping Yu, University of Chicago

Organic Display Materials	Norman Bardsley, United States Display Consortium
Membranes	William Koros, University of Texas-Austin

Bioderived and Bioinspired Materials

Bioderived and Bioinspired Materials—Overview	Ilhan Aksay, Princeton University
Bioderived and Bioinspired Functional Materials	Mark Reed, Yale University
Engineering Cell/Material Interfaces	Carolyn Bertozzi, University of California, Berkeley
Memory Implants for Human Performance	Theodore Berger, University of Southern California
Computational Theory and Modeling	Alan Windle, University of Cambridge

October 11, 2001

Structural and Multifunctional Materials

Polymer/Nanotube Composites	Larry Drzal, Michigan State University

Energy and Power Materials

Power Systems (Marine Fuel Cells)	John Sofia, NAVSSES
Microelectromechanical Systems	William Tang, DARPA
Computational Materials Science as It Relates to Energy/Power	Brian Wirth, LLNL

Electronic and Photonic Materials

Magnetic Materials	Ivan Schuller, University of California, San Diego

October 12, 2001

Electronic and Photonic Materials

Packaging Materials	Charles Bauer, TechLead Corp

Functional Organic and Hybrid Materials

Photovoltaics	Yang Yang, University of California, Los Angeles

JANUARY 29, 2002
NATIONAL RESEARCH COUNCIL
WASHINGTON, D.C.

January 29, 2002

Panel on Bioderived and Bioinspired Materials	Michael Jaffe
Panel on Functional Organic and Hybrid Materials	Frank Karasz
Panel on Electronic and Photonic Materials	Julia Phillips
Panel on Energy and Power Materials	James Baskerville
Panel on Structural and Multifunctional Materials	Harry Lipsitt
DoD Vision and System Needs—Update	Andrew Marshall, OSD

APPENDIX B

Biographical Sketches of Committee and Panel Members

COMMITTEE ON MATERIALS RESEARCH FOR DEFENSE AFTER NEXT

Harvey Schadler, *chair* (NAE), is a retired technical director of the General Electric Corporate Research and Development Center. He was elected to the NAE for exceptional leadership in the development and application of advanced materials and processes in the electrical and aircraft engine industries. His expertise is in the physical properties and processes of manufacture of magnetic, superconducting, high-temperature, and nuclear metallic and ceramic materials. His expertise includes aerospace and Army systems.

Alan Lovelace, *vice chair* (NAE), is a retired senior corporate vice president and chairman of Commercial Launch Services, General Dynamics Corporation. His expertise includes aerospace and defense systems and materials. He was elected to the NAE for his contributions to aerospace materials, particularly the application of boron- and graphite-reinforced epoxy composites.

James Baskerville is vice president for surface ship support and for advanced technology at Bath Iron Works (General Dynamics). His expertise is in Navy systems and materials. He joined Bath Iron Works in 1997 after serving more than 25 years in the U.S. Navy. He is a registered professional engineer with extensive experience in the development and use of marine composites.

Federico Capasso (NAS, NAE) is vice president of the Physical Research Laboratory, Bell Labs, Lucent Technologies. His expertise is in the area of electronic materials. He pioneered the use of bandgap engineering as a powerful tool in the design of semiconductor devices and heterostructures and made related seminal contributions to electronics, photonics, and semiconductor science in the areas of detectors, lasers, transistors, quantum devices and circuits, and artificial structures with new transport and optical properties.

Millard Firebaugh (NAE), a retired U.S. Navy rear admiral, is vice president of innovation and chief engineer for the Electric Boat Corporation (General Dynamics). His expertise is in naval systems and materials, submarine design, and naval architecture.

John Gassner is director of the Supporting Science and Technology Directorate and chief scientist of the U.S. Army's Natick Soldier Center. His expertise is in polymer science and engineering, composites development, manufacturing processes, and Army systems. His research has included work on high-performance polymers, nanomaterials, self-assembly, sensors and smart structures, advanced materials for energy attenuation, and innovative processing techniques for organic materials.

Michael Jaffe is a faculty member at the New Jersey Institute of Technology and Rutgers University. He is chief scientist for applied programs and director of the Medical Device Concept Laboratory of the New Jersey Center for Biomaterials and Medical Devices, Rutgers. His expertise is in innovative materials research areas, such as biomimetics, the structure-property relationships of polymers and related materials, the application of biological paradigms to materials design, and the translation of new technologies to commercial reality.

Frank Karasz (NAE) is Silvio O. Conte Distinguished Professor, Department of Polymer Science and Engineering, University of Massachusetts. His research activities are concentrated in polymer physics and chemistry: polymer-polymer interactions in binary amorphous and amorphous crystalline blend systems; effects of copolymerization and microstructure; nuclear magnetic resonance studies of polymer solid state, especially blends; computer simulations of polymer-polymer miscibility; quasi-elastic light scattering from macromolecular solutions; electronic and optical properties of conducting polymers; and polyimide systems.

Harry A. Lipsitt is professor emeritus in the Department of Mechanical and Materials Engineering at Wright State University. His expertise is in intermetallics and metals. He spent 30 years at the Air Force Wright Laboratories working on the development and optimization of metallic and intermetallic materials for use in high-temperature applications.

Meyya Meyyappan is director of the Center for Nanotechnology at the NASA Ames Research Center. His expertise is in nanotechnology, carbon nanotubes and sensor materials, and electronic materials. At NASA, he is responsible for basic research in nanotechnology, computational semiconductor device physics, computational and experimental chemistry in materials processing, and process/equipment modeling.

George Peterson is a retired director of the Materials Laboratory at the U.S. Air Force Wright Aeronautical Laboratories. His expertise is in the properties and processing of structural polymer composites, manufacturing technologies, and aerospace systems. Under his direction, the Air Force focused its efforts on developing low-cost production processes for electronics, nonmetallics, and metallics.

Julia M. Phillips is director of the Physical and Chemical Sciences Center at Sandia National Laboratories. Her expertise is in electronic materials, materials characterization, and computational materials science. Her accomplishments have been in the areas of epitaxial metallic and insulating films on semiconductors; high-temperature superconducting, ferroelectric, and magnetic oxide thin films; and novel transparent conducing materials.

Richard Tressler is professor emeritus of the Department of Materials Science and Engineering at Pennsylvania State University. His expertise is in the properties and processing of structural ceramics and ceramic composites, as well as Army and aerospace systems. His research interests include the fabrication and mechanical behavior of structural ceramics, ceramic composites, fracture and strengthening mechanisms, and the correlation of processing with ceramic properties.

PANEL ON STRUCTURAL AND MULTIFUNCTIONAL MATERIALS

Harry A. Lipsitt, *chair* (see committee biographical sketch).

Millard Firebaugh (NAE), *vice chair* (see committee biographical sketch).

Michael I. Baskes is a staff member at Los Alamos National Laboratory in the Structure/Property Relations Group. His interests encompass the use of computational methods to investigate material properties such as alloy phase stability, magnetic behavior, fracture toughness, empirical and semi-empirical potential development for metals and semiconductors, diffusion and trapping of hydrogen isotopes, interfaces and grain boundaries, atomistic calculations of dislocations in metals, brittle and ductile fracture, and microsegregation in welding. He established the journal *Modeling and Simulation in Materials Science and Engineering* in 1992, and has been its editor-in-chief since its inception.

L. Catherine Brinson is an associate professor in mechanical engineering, Northwestern University. Her research interests include constitutive modeling of shape memory alloys and mechanical behavior of other smart materials; research on time dependence of polymeric properties at scales ranging from the nanoscale; microscale interlayer effects; macroscopic thermomechanical response; and recently microporous metallic alloys with emphasis on biological applications. Research projects often focus on environmental effects on material behavior, such as material aging and coupled temperature and mechanical response.

Thomas W. Eagar (NAE) is the Thomas Lord Professor of Materials Engineering and Engineering Systems, Massachusetts Institute of Technology. He has made many contributions to the theory and practice of welding. At MIT he has also served as director of the Materials Processing Center and codirector of the Leaders for Manufacturing Program. He has served on technical committees for U.S. governmental departments and agencies and has held numerous officer positions in many professional associations.

Richard J. Farris is Distinguished University Professor, Department of Polymer Science and Engineering, University of Massachusetts. His research has had a pronounced effect on how polymer properties are determined and modeled. His teaching and research have focused on development of a fundamental understanding of the physical and mechanical behavior of polymeric materials using the classical continuum methods of experimental mechanics, physics, and thermodynamics. His research group is recognized as the source of many unique instruments and characterization techniques for probing important but difficult-to-measure phenomena. He specializes in exploring the behavior of solid polymers and composites, with a strong emphasis on fibers and coatings.

D. David Newlin is director of vehicle survivability technologies, General Dynamics Land Systems. He is responsible for detection avoidance, hit avoidance, penetration avoidance, and kill avoidance technologies, including armor, signature management, and advanced materials and structures. He previously served 26 years in the Army in a variety of positions, including product manager for the common chassis and project manager for ground system integration, which worked on ground vehicle survivability technology investment.

George Peterson (see committee biographical sketch).

Richard Tressler (see committee biographical sketch).

PANEL ON ENERGY AND POWER MATERIALS

John Gassner, *co-chair* (see committee biographical sketch).

James Baskerville, *co-chair* (see committee biographical sketch).

Daniel H. Doughty is manager of the Lithium Battery Research and Development Department at the Sandia National Laboratories. The department is responsible for developing advanced power sources, typically batteries and electrochemical cells based on lithium. It also works on leading edge electrochemistry as well as advanced batteries and battery materials for defense and commercial applications. Dr. Doughty previously worked as a research chemist for 3M Co. developing advanced inorganic photoconductors. His areas of expertise include lithium chemistries (e.g., lithium ion rechargeable batteries and lithium thionyl chloride cells and batteries) and general materials chemistry and processing, including colloid chemistry, superconducting ceramics, intercalation compounds, and oxide surface chemistry.

Sossina M. Haile is associate professor of materials science, California Institute of Technology. Her research centers on ionic conduction in solids, with the twin objectives of understanding the mechanisms that govern ion transport and applying that understanding to development of advanced solid electrolytes and novel solid-state electrochemical devices. Technological applications of fast ion conductors include batteries, sensors, ion pumps, and fuel cells. It is in this last area that Dr. Haile's work is expected to have the most impact.

Robert N. Katz is the Norton Research Professor of Materials Science and Engineering, Worcester Polytechnic Institute, and the principal of R. Nathan Katz Associates. He was chief technologist at the Army Research Laboratory, Materials Directorate, at Watertown, Massachusetts, from 1987 through 1995 and was chief of the Ceramics Research Division there from 1970 through 1987. He is an internationally recognized expert in advanced ceramics and ceramic composites and their application in areas of both military and civilian importance, including ceramic composite armor, missile guidance transparencies, cutting tools, bearings, and most especially heat engines. His research interests include structure/property relationships at high temperature in high-performance ceramics, high strain rate properties of ceramics, tensile testing of ceramics and ceramic matrix composites, design with brittle materials, and technology assessment.

PANEL ON ELECTRONIC AND PHOTONIC MATERIALS

Julia M. Phillips, *co-chair* (see committee biographical sketch).

Meyya Meyyappan, *co-chair* (see committee biographical sketch).

Harold G. Craighead is Charles W. Lake Professor of Engineering and Interim Dean of the College of Engineering at Cornell University. His research centers on the new science and applications of nanometer-scale devices and structures with special attention to nanofabrication processes and their impact on the properties of materials and devices. He is also working to advance understanding and manipulation of the physical properties of systems of reduced dimensions. Current research topics include high-resolution electron-beam processes and the physics of ultrasmall structures. He is investigating application of these in the fields of optics, magnetism, and biology.

Narsingh B. Singh is manager of crystal science and technology at Northrop Grumman Corporation. His expertise is in crystal growth of optoelectronic materials and materials for novel electro-optical devices. His research interests include materials for nonlinear optical, acousto-optical, and x-ray and gamma-ray detector applications and single crystals grown by the physical vapor transport, solution growth, and Bridgman crystal growth methods. He has also studied electronic and photonic materials for lasers, acousto-optic-based sensors, RF materials by LPE for frequency controls, dielectric materials for phase shifters, periodically poled thick films for mid-IR lasers, and AlN substrates for GaN epitaxy.

Ming C. Wu is a professor, Department of Electrical Engineering, and director, MURI Center on RF Photonic Materials and Devices, at the University of California at Los Angeles. His expertise is in micromachined microoptics, optical microelectromechanical systems, free-space integrated optics, high-speed optoelectronics, microwave photonics, high-power photodetectors, and mode-locked semiconductor lasers. He has also studied high-speed semiconductor lasers and optoelectronics.

Edward Zellers is a professor in the Departments of Environmental Health Sciences and Chemistry and group leader in the NSF Engineering Research Center for Wireless Integrated MicroSystems, University of Michigan. His areas of expertise include the design and implementation of microfabricated chemical sensors and sensor arrays for measurement of small organic molecules; novel neutral and conducting polymers and nanostructures as chemical sensor interfaces; microanalytical systems for organic analytes in air and biological media; polymer characterization and testing in sensor systems and as permeation barriers; exposure assessment strategies; and analytical chemical methods and instrumentation.

PANEL ON FUNCTIONAL ORGANIC AND HYBRID MATERIALS

Frank Karasz (NAE), *chair* (see committee biographical sketch).

Lisa Klein is a professor in the Department of Ceramic and Materials Science and Engineering at Rutgers, the State University of New Jersey. Her expertise is in ceramics and ceramics processing, particularly sol-gel technology. She has been at Rutgers University since 1977, after completing postdoctoral research at Massachusetts Institute of Technology. In recent years, she has worked in a visiting professor capacity with Sandia National Laboratories/USA, Laboratoire D'Energetique Electrochimique/France, and the Hebrew University of Jerusalem/Israel.

Vincent D. McGinniss is senior vice president, Materials Sciences, at Optimer Photonics, Columbus, Ohio. His expertise covers polymer chemistry, specialty coatings, specialty adhesives, fiber-reinforced composites, photoactive catalyst systems, water-based polymers or emulsions, fluorine chemistry, electrodepositable polymer materials, low-temperature cure reactions for coatings and adhesives, and radiation-curable coatings and technology. He is a leading authority on specialty chemical and polymer systems for a wide range of applications, having devised many novel photoinitiator systems, composite systems, water-based coatings, and specialty coatings for commercial use in the wood, can, and floor covering industries.

Gary E. Wnek is professor and chair, Department of Chemical Engineering, and affiliate professor of chemistry and biomedical engineering at Virginia Commonwealth University and adjunct professor of chemistry at Virginia Tech. His research deals with polymers with unusual electrical or optical properties, electric field-modulated phenomena, biosensors and biochips, and artificial axons. He is a member of the Editorial Board for *Polymer-Plastics Technology and Engineering* and has served on the editorial boards for *Chemistry of Materials* and *Progress in Polymer Science*.

Luping Yu is a professor in the Department of Chemistry at the University of Chicago. His expertise is in the rational design, synthesis, and characterization of novel functional and multifunctional polymers and molecules, paying great attention to new polymerization methodologies for the synthesis of these new materials. Typical current projects are the development of polymerization methodologies; design, synthesis, and physical studies of photorefractive polymers; second-order nonlinear optical polymers; conjugated diblock polymers for supramolecular self-assembly of nanostructured materials; architectural controls of polymer microstructures, including synthesis of dendritic polyphenylenevinylenes; metal-containing functional polymers; and supramolecular functional materials.

PANEL ON BIODERIVED AND BIOINSPIRED MATERIALS

Michael Jaffe, *chair* (see committee biographical sketch).

Ilhan Aksay is a professor in the Department of Chemical Engineering and the Princeton Materials Institute of Princeton University. He provides expertise in bioceramics: His current research is on the utilization of colloidal and biomimetic techniques in ceramic processing. In recent years, his work has mainly focused on the utilization of complex fluids to control the architecture of organic/ceramic nanocomposites. His research has resulted not only in contributions to the literature on the fundamentals of ceramic processing but also in products produced by the industry.

Mark Alper is deputy director, Materials Sciences Division, Lawrence Berkeley National Laboratory (LBNL), and adjunct professor, Department of Molecular and Cell Biology, University of California at Berkeley. His expertise is in biochemistry and materials science. He has held a number of other research management positions at LBNL, including executive director, Chancellor's Biotechnology Council; associate division head, Materials Sciences Division; and deputy director, Center for Advanced Materials. Dr. Alper founded the LBNL Biomolecular Materials Program and has been director since its inception.

Paul Calvert is a professor, Department of Materials Science and Engineering, University of Arizona. His recent research has revolved around his area of expertise, biomimetic materials; he is studying methods of forming composites by in situ mineralization to mimic bone and tooth, layerwise freeform fabrication to parallel biological growth, methods for producing implant materials, and inkjet printing as a method for building structures from biological and synthetic polymers. Other research areas of interest include polymer-ceramic composites, conducting polymers, diffusion and permeability in polymers, composite interfaces, polymer crystallization and crystallization kinetics, composite materials for bone prostheses, carbon nanotube composites, and free-formed metal-ceramic multilayers.

Mauro Ferrari is a professor of internal medicine, of mechanical engineering, and of materials science; director, Biomedical Engineering Center; and associate director, Dorothy M. Davis Heart and Lung Research Institute, all at Ohio State University. His expertise is in medicine, biomechanics, biotransport phenomena, drug delivery, and biomedical micro/nanotechnology, and his research interests include biomedical nanotechnology and bioMEMS for drug delivery, cell transplantation, and tissue engineering, with emphasis on oncology, cardiovascular disease and diabetes; biosensors and bioseperation; multiscale discrete/continuum mechanics and biomechanics; mechanics of composite materials and structures; thermoelasticity; homogenization theory; and functionally graded materials.

Erik Viirre is an assistant adjunct professor, Division of Otolaryngology, University of California at San Diego School of Medicine, and a consultant to laboratories and industries. His research and clinical interests cover vision, hearing, and the balance-sensing system of the body called the vestibular apparatus. His clinical specialties are diagnosis and treatment of vertigo, motion sickness, and balance problems, as well as tinnitus. His recent research has dealt with the use of virtual reality technology in rehabilitating patients with balance disorders, vision research on an advanced visual display technology called the virtual retinal display, and the development of a virtual clinic where he will see patients in remote locations using telemedicine. He has already demonstrated the diagnosis and treatment of vertigo patients in remote locations.

Integration of Materials Systems and Structures Development

Forty years ago, Westbrook[1] noted that structural materials vary in price by seven orders of magnitude, from gravel or cement at several cents per pound to industrial diamonds at \$10,000 per pound. Not surprisingly, usage of a material in pounds per annum is inversely related to its cost per pound (see Figure C-1). However, Westbrook's graph reveals an interesting principle: Judging from the slope of the usage cost trend as compared to the lines of equal market size, a significant reduction in the cost of a material will result in a larger increase in usage. For example, a reduction of a factor of two in the cost of a material should result, over the long run, in a fourfold increase in usage. Thus, one way to increase the size of the market for a structural material is to reduce the cost.

It is also useful to consider the value of a pound of weight saved over the life of a vehicle (see Table 3-2). With gasoline at \$1 to \$2 per gallon, a pound of weight removed from an automobile will save \$2.00 over a 100,000-mile life. For a commercial aircraft, the fuel savings over a 100,000-hour life of the fuselage is \$200 per pound. For military aircraft, the value can be \$1,000 per pound. For spacecraft, the cost to put a pound of payload into orbit a single time is \$20,000; for the reusable space shuttle the cost can drop to \$10,000 per pound. The goal for a single stage to orbit shuttle is \$1,000 per pound, but this has not yet been achieved.

---

[1]Westbrook, J.H., Internal General Electric Report, General Electric, Schenectady, NY, 1962.

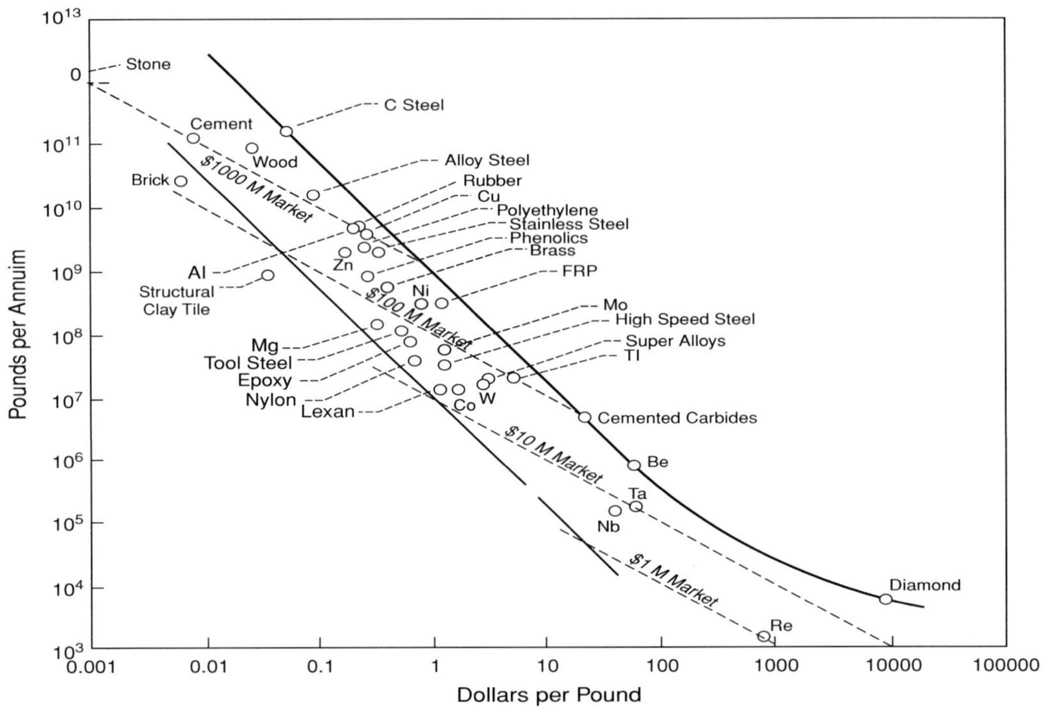

FIGURE C-1

Price-volume relationship for annual U.S. consumption of structural materials. SOURCE: J.H. Westbrook, General Electric (retired), private communication, September 27, 2002.

In spite of concerns about the cost of raw materials, their cost is only a relatively small fraction of the cost of a fabricated structure, typically 10 to 20 percent (see Table C-1). Combining the fabricated costs of a structure with the value of a pound saved gives maximum average cost of the material in a particular application. For example, for an automobile where the value of a pound saved is $2, $2 times a 20 percent cost of material in relation to total cost produces an upper limit (on average) of $0.40 per pound for the primary structural material of the automobile. Automotive-quality steel sheet is $0.30 per pound, while aluminum sheet is $1.50 per pound. Thus, even with a 250 percent lower density, aluminum cannot be justified in current automobiles when gasoline is $1.50 per gallon; esti-

TABLE C-1 Typical Costs of a Fabricated Structure Made from Monolithic (Noncomposite) Materials

Component	Percentage of Total Cost
Raw Materials	10-20
Design/Engineering	10-20
Fabrication (forging, machining, joining, etc.)	20-40
Nondestructive Testing and Quality Control	10-20
General and Administrative	10
Profit	+10 to −10

mates of the breakeven point for aluminum auto bodies is $4.00 per gallon.[2]

Combining the values of Table 3-2 with the costs shown in Table C-1 makes it possible to generate a list of materials for various structures (see Table C-2). It should be noted that complex composites that possess remarkable mechanical properties can be equally remarkable in cost. For example, the liquid hydrogen tank for the X-33 space plane, one of the largest complex composite structures ever built, was the size of a small house yet weighed only 4,000 pounds. However, the fabricated cost per pound was $10,000. Thus, the percentages in Table C-1 do not apply to complex composites. For such composite structures, the material costs may be as little as 2 to 5 percent of the total fabricated structure cost.

Note that while light weight is important for anything that moves, the faster the object moves, the greater the value of weight saved. Thus, although the average value of a pound saved in an automobile is $2.00, the savings on reduced weight in an axle or wheel can be double or triple this value since the wheels rotate faster than the main structure—that is why aluminum wheels are cost-competitive with steel in automobiles. In an aircraft, a pound of weight saved on a disk of a turbine engine can be worth 10 times the same weight saved on the fuselage, because a pound saved on the engine can save 5 to 10 pounds on the wing structure.

While these values of weight saved and materials costs as a fraction of total structure costs may appear rather general, experience has shown these estimates to be surprisingly accurate. For example, Newport News Shipbuilding has estimated $10,000 per ton savings if a higher strength

[2]P. Bridenbaugh, Alcoa, private communication, October 11, 2001.

TABLE C-2 Structural Materials Selection Based on Value of Weight Savings over the Life of a Structure

	Dollars per Pound
Automobiles, ships, and buildings	
Reinforced concrete	0.15
Cast iron	0.15
Mild steel	0.15
Low alloy steel	0.25-0.75
Plywood	0.40
Aircraft	
Polyethylene	0.65
Hardwoods	0.70
Rubber	0.75
Glass	0.75
Epoxy	0.85
Aluminum	1.00-1.50
Polycarbonate	1.25
Copper	1.50
Stainless steels	1.50-3.00
Graphite fiber-reinforced plastic	1.25-1.75
Nickel alloys	3.50-15.00
Titanium alloys	5.00-15.00
Cobalt	10.00-30.00
Boron-epoxy composites	150.00
Spacecraft	
Complex composites	100.00-500.00
Refractory metals	100.00-300.00
Silver	150.00
Gold	5,000.00

ship plate (HSLA 65) can be substituted for the former steel.[3] Since the higher strength steel would cost approximately $0.50 per pound, at 10 percent material, compared to total fabricated cost, the estimated savings aligns perfectly with these general rules.

[3]P.J. McMullen, "Optimized HSLA-65 Welding Procedures for Fabrication of Naval Ship Structures," National Center for Excellence in Metalworking Technologies, Concurrent Technologies Corporation, Johnstown, PA, TR-No. 97-176.

APPENDIX D

Energy and Power Materials

A NOTE ON POWER SOURCES

There is a continuing need for materials that can store large amounts of energy in small packages and use the energy efficiently. In materials terms, advanced energy storage requires materials that have high volumetric and gravimetric energy densities. While many of the approaches discussed in this report can affect several types of power sources, it is important to realize that no single power source is appropriate for all applications (see Figure D-1).

A range of energy storage and conversion technologies will have central roles in engineered energy storage and delivery systems for powering DoD needs. *High-energy* needs are likely to be met with a liquid fuel because the energy density of liquid fuels, in reaction with air in an internal combustion device or fuel cell, is unsurpassed. *High-power* applications, on the other hand, are likely to require batteries, or more likely, hybrid power devices that combine a high-energy system (e.g., a fuel cell) with a high-power system (e.g., a powerful battery or electrochemical capacitor). Properly designed, a system can enjoy the benefits of both the high energy content of a liquid fuel and the high power of a battery.

In the main body of this report, the committee attempts to show that some material improvements can affect several technologies. Batteries and fuel cells will both benefit from improved electrolytes and better methods to tailor electrochemical interfaces. Other improvements will affect one technology more than another or may even be specific to a single technology.

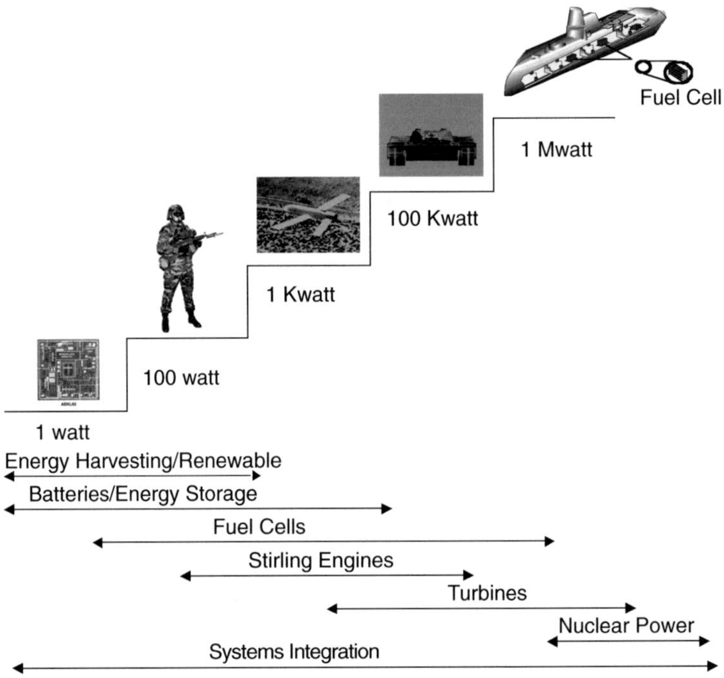

FIGURE D-1

Military systems power requirements often follow a "step function," so different power sources are needed for different applications. SOURCE: Carlin, Richard, ONR, Electric Power Sources for the Navy and Marine Corps, Presented at the Navy Grand Challenges Workshop, November 16-18, 1999.

COMMERCIAL VERSUS MILITARY BATTERY REQUIREMENTS

Many specific military requirements limit the use of commercial off-the-shelf (COTS) power sources. These include differences in required power and energy levels, in temperature operating range, in the need for shock resistance, in longevity requirements, and in reliability levels. Except in hybrid electric vehicles (HEVs), high-power batteries are generally not needed in the commercial marketplace, and even HEVs are not subject to the same performance demands that future defense applications will encounter. Materials for military use must also operate in a broader temperature range and under more rigorous environmental extremes than materials for commercial use. Military batteries must have a longer shelf life and lower self-discharge rates than typical commercial batteries.

Furthermore, commercial devices are not held to the same reliability standards as military devices. The reliability driver for commercial power sources is to minimize warranty expenses; the reliability driver for future defense systems is mission success and soldiers' lives. These differences, which are present in all the Services, are typified in the requirements for Navy batteries shown in Figure D-2.

DIELECTRIC ENERGY STORAGE

Capacitor dielectrics will continue to have an important role for DoD, particularly for energy storage, power conditioning, high-rate switching, and pulsed power for a range of future DoD systems. Compared to other forms of electrical energy storage, capacitors are lower in energy density

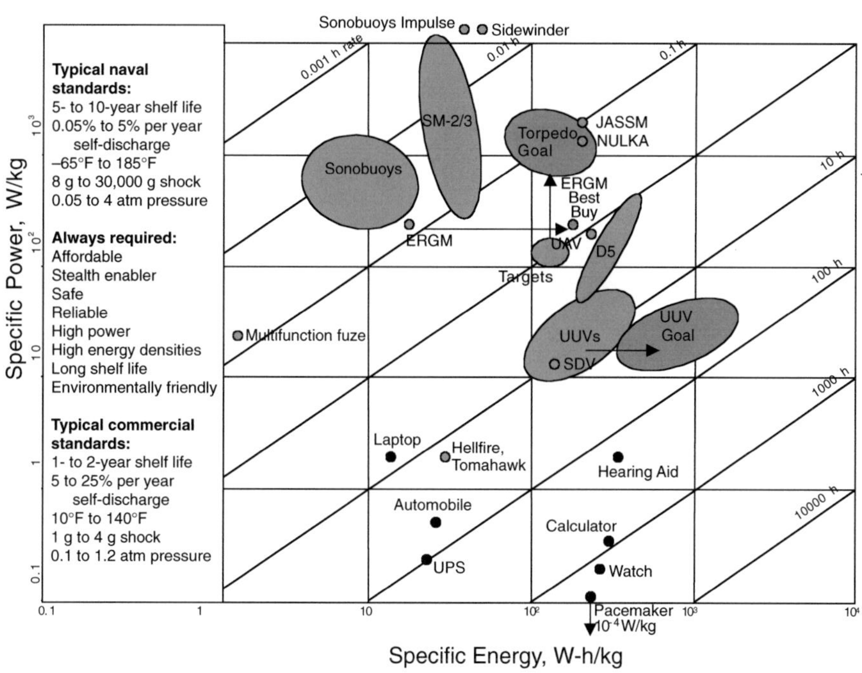

FIGURE D-2

Military versus commercial requirements for batteries. SOURCE: Suddeth, David, Naval Surface Warfare Center, Managing Power Source Strategies for Navy Applications, Presented at Navy Conference in Crystal City, VA, December 11-13, 2001.

but higher in power density (see Figure D-3), and they are more readily cycled at high frequencies.

EXPLOSIVES AND PROPELLANTS

Explosives

Many of the key chemical explosives of military interest today were first synthesized in the late 1800s and, as with structural materials, moving from laboratory curiosity to general application of energetic materials has often taken 20 to 30 years, as shown in Table D-1 (Federoff, 1960). Important technological parameters for high explosives are energy release/reaction propagation rate; energy density; and resistance to accidental explosion (insensitivity). Increased insensitivity to accidental detonation

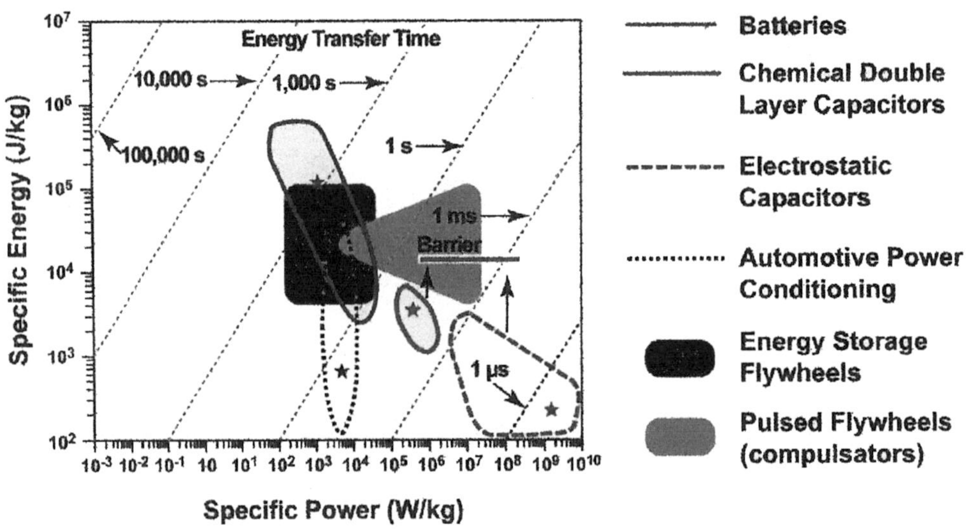

FIGURE D-3

Power versus energy density for selected mechanisms for electrical energy storage. SOURCE: Clelland, I., R. Price, and J. Sarjeant. 2000. Advances in Capacitor Technology for Modern Power Electronics. Pp. 145-148 in Proceedings of the 24th International Power Modulator Symposium, Norfolk, VA, IEEE, Piscataway, NJ.

TABLE D-1 First Synthesis of Chemical Explosives of Military Interest

Material	Year First Reported or Synthesized	Year First Used in Large-Scale Service
Nitroguanidine	1877	
Tetryl	1879	1906
TNT	1880	1900
TATB	1888	
PETN	1894	1930
RDX	1899	1935
Picric acid	1900	1910
HMX	1930	1970

NOTE: HMX was used as early as the 1950s in explosive compositions but only reached widespread use in about 1970.

has been a major focus of modern explosives research over the past 40 years, and efforts continue to reduce sensitivity while increasing explosive energy.

Over the last 50 years, the average power of military high explosives has increased by approximately 40 percent. All indications are that tomorrow's weapon systems will be smaller than today's. Unless they are to carry a disproportionate share of mass in weapons and explosive power, achieving the same energy on target will require higher energy density materials (HEDM).

Is there room for additional improvement? Figure D-4 depicts the mass-based energy density spectrum (cal/gm) and the relative position of conventional high-energy chemistry within it. Conventional hydrocarbons based on oxygen and nitrogen (CHNO) explosive materials are at the low end, on the order of 10^3 cal/gm. New CHNO-based molecules combined with advanced approaches for achieving high surface areas and mixing components may be able to increase this figure several-fold—the equivalent of several centuries of progress. The figure indicates that there may be even much greater potential for HEDM based on alternative approaches. Some of the latter are currently only at the conceptual stage; others are just entering the basic research stage.

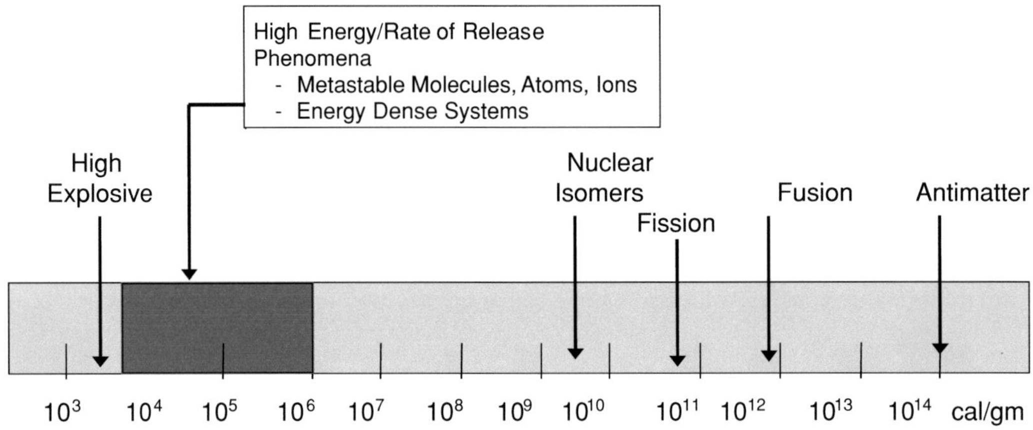

FIGURE D-4

Is there room for improvement for energetic materials? Energy density per unit mass. SOURCE: Ullrich, G.W., director, Weapons Systems, Office of the Secretary of Defense, Advanced Energetic Materials: Introduction and Overview, presented to the Committee on Advanced Energetic Materials and Manufacturing Technologies, National Research Council, Washington, DC, July 31, 2001.

Propellants

Propellants burn without exploding and contain all the oxygen they need for combustion (Alchavan, 1998). Ideal propellants are dense; occupy little space before burning; produce only low-molecular-weight gases and heat; show repeatable performance; yield acceptably low gas temperature and products to reduce gun damage; provide maximum gas output within the pressure limits of the gun; resist shock, impact, and heat; are safe to store and handle; and are low-cost and environmentally acceptable. Specific impulse (I_{sp}) is one of the performance measures most often used for propellants. An additional measure of efficacy is the amount of oxygen needed to convert all oxygen to carbon dioxide and all hydrogen to water (Cooper and Kurowski, 1996). Propellants emit gas and heat but, in contrast to explosives, do not react faster than the bulk sound speed in the material.

Traditional propellants (single-, double-, or triple-base materials) are composed of CHNO molecules that self-oxidize; they have historically been used in guns or small military rocket motors. In a separate class of materials is the composite propellant, which contains fuel and oxidizer as separate components, and which is used extensively in rockets and for gas generators other than in guns. The fuel often acts as a binder for the components. Although both liquids and solids have been employed as rocket/missile propellants,[1] all major gun propellants today are solid.[2] This report concentrates on opportunities in solid propellants because of their widespread use across DoD and their similarities to explosives.

Key Issues for Energetic Materials

Much effort over the past few decades has been directed to safety and process improvements in explosives and propellants. In particular, creating munitions insensitive to accidental detonation has been a major goal. Traditional approaches to development of insensitive materials for explosives and propellants have relied quite heavily on intuition. This process is time-consuming, with no guarantee of success. To date, there has been relatively little application of atomistic modeling to energetic materials formulation. The problem has been recognized by issuance of a Defense Strategic Research Objective (SRO) on Insensitive High-Energy Materials.[3] A new paradigm is slowly evolving in which science-based models are supplementing, and in some cases replacing, intuitive approaches to munitions design. One estimate is that development time for propellants using current ingredients requires 10-15 years, whereas development of new materials and the associated formulations can require 25-40 years,[4] in part because empirical testing requires use of multiple iterations at multiple scales. An important goal is to reduce this development time.

[1]Hunley, J.D. "AIAA Invited Paper—The History of Solid-Propellant Rocketry: What We Do and Do Not Know," paper presented at the 35th AIAA, ASME, SAE, ASEE Joint Propulsion Conference and Exhibit, Los Angeles, CA, June 20-23, 1999.

[2]Forch, B.E., "Energetic Materials for the Objective Force," paper presented by B.E. Forch, Army Research Laboratory, at the National Defense University, Ft. Leslie J. McNair, Washington, DC, July 9, 2002.

[3]Forch, Brad E., and Betsy M. Rice, "Strategic Research Objective: Insensitive High-Energy Materials," briefing presented to the Panel on Energy and Power Materials of the Committee on Materials Research for Defense After Next, National Research Council, Washington, DC, March 29, 2001.

[4]Goldwasser, J., "Navy Energetic Materials Science and Technology Programs," briefing presented to the Committee on Advanced Energetic Materials and Manufacturing Technologies, National Research Council, Washington, DC, July 31, 2001.

The development of novel energetic materials over time has relied on the intuition, ingenuity, and knowledge of experienced individuals and on extrapolation of the properties and structures of existing energetic materials to those of new materials. These complex materials will continue to be a fundamental contributor to U.S. national defense in the future.

To streamline the R&D process in this area, effort should be directed at assembling and integrating the entire arsenal of materials development capabilities. These include use of (1) computation materials science in materials design to help identify promising new molecules and formulations; (2) new material synthesis approaches; (3) improved processing methods; (4) new characterization techniques; and (5) advanced modeling to reduce the expensive and time-intensive testing necessary to verify performance.

Finding advanced approaches to computation materials science, combining first principles and known design rules with the results of well-designed experiments, has the potential for reducing the amount of time required to identify, develop, and verify the performance of new energetic materials. Already, molecular modeling techniques applied to the development of new energetic materials are showing promise of materials that provide enhanced energy density and insensitivity.

It is important to understand the fundamental processes for initiation of energetic materials and, in the case of propellants, the associated fundamental combustion mechanisms. Having a comprehensive model of reaction processes in heterogeneous nanoscale energetic materials, taking into account details of initiation and propagation of reaction fronts, would help us understand these complex processes and would form a basis for design and development of novel energetic materials.[5]

Many of the materials that have already been identified as promising have no known synthesis routes. Others can be prepared, but only in gram quantities by costly means requiring either multistep synthetic approaches, expensive reagents, difficult-to-maintain experimental conditions, or combinations thereof. Making novel energetic materials available in large quantities for military use will require improved means for synthesis and

[5]Wilson, W.H., munitions directorate, Air Force Research Laboratory, Eglin Air Force Base, "Advanced Energetic Materials Research," paper presented to the Committee on Advanced Energetic Materials and Manufacturing Technologies, National Research Council, Washington, DC, July 31, 2001.

new processes for combining these materials with others in complex formulations.

As is discussed below, one promising avenue to new energetic explosives and propellants is to use nanomaterials technology and nanocomposite materials. In parallel with developments in the rest of this emerging field, this will require new techniques for characterizing these materials at different size scales, particularly the nanoscale. Of particular interest in these nanocomposite energetic materials are such factors as nanoparticle size, size distribution, morphology, surface chemistry, and composition. These variables have all been shown to be important in determining mechanical, barrier, and other properties of nanocomposites (Kornmann et al., 1998; Messersmith and Giannelis, 1995), and they will likely control the contributions of nanoscale components in energetic materials as well.

Finally, the time required to field new energetic materials is long, due in no small part to the rigorous testing required to assure their performance, reliability, and safety. Again, use of new computational approaches may help compress this time schedule.

LOGISTIC FUELS

An overriding military concern in fuels technology is logistics. Fewer fuels mean a simplified logistics burden and the ability to provide energy, rapidly and flexibly, to the system that most requires it. From this perspective, all military systems would ideally run on a single fuel.

Many current military propulsion systems have been designed to use a variety of petroleum-based fossil fuels, such as jet fuel (e.g., JP-5, JP-8) and diesel. These fuel families are exceedingly well-established in military applications and are supported by a strong logistics infrastructure. The panel recognizes that new electrochemical propulsion and power systems employing fuel cells are in various stages of design and development and it believes that their development as outlined in Chapter 4 is essential for the DoD applications of 2020.

However, it also appears that current hydrocarbon military fuel families (jet fuel and diesel fuel) will continue to power most military platforms in 2020 due to their established position as workhorse fuels, combined with the cost and time required for propulsion system redesign to produce other fuels. Convergence to use of a single petroleum-based logistic fuel is also unlikely because diesel fuel, now used in a variety of trucks and military ground vehicles, will not flow readily in a jet engine, while gasoline is too volatile. Therefore, while there is a strong push toward novel fuels for fuel cell applications, in 2020 fuels for turbine and internal combustion engines used to transport men and materiel are expected to remain primarily hydrocarbon-based.[6] Reformed diesel and jet fuels are also likely to retain a prominent role even when fuel-cell-based systems arrive in large numbers, due to the strong established base for these fuels.

These conclusions are supported by examination of fuel-related developments in the commercial sector, which shows a continued push to improve petroleum-based fuels and power sources. In 2020, oil and natural gas are expected to remain the dominant energy supply for the United States, with civilian transportation fuels obtained primarily from hydrocarbons. Production of petroleum-based fuels is expected to peak between 2020 and 2040. Throughout the transportation sector a global shift to cleaner fuels is expected by 2020, raising energy efficiency and reducing greenhouse gas emissions. Concurrent efforts will be made to improve civilian transport efficiency by way of leaner burning internal combustion systems, hybrid electric vehicles, fuel cells, greater use of mass transportation, and telecommuting. The environment will drive changes in petroleum-based fuel formulation to reduce volatile organic compounds, NO_x, particulates, and greenhouse gas emissions. Most industrialized nations will continue to seek reductions in SO_x emissions in jet fuels while retaining fuel lubricity and performance. In the civilian sector, hybrid internal combustion/electric drives and other novel drive systems will evolve for automotive engines. Power sources in civilian automotive applications will include both petroleum fuels for Carnot cycle engines and hybrid gas-electric systems, with portions of the market being filled by approaches such as all-electric power, compressed or liquefied natural gas, or hydrogen. Because these latter approaches will need considerable new infrastructure to be effective, their widespread adoption is likely to be delayed.

Improvements to today's logistic fuel technology are expected to be largely evolutionary. While many of the problems are not materials-related

[6]Gautam, R., "Fuels Processing," briefing to the Panel on Energy and Power Materials of the Committee on Materials Research for Defense After Next, National Research Council, Irvine, CA, October 10, 2001.

per se, their solutions could lead to improvements in military fuels. Modifying refinery yield to improve fuel quality (e.g., by reducing sulfur) or to optimize yield of one particular fuel fraction is one such evolutionary approach. Another is fuel interconversion using ring-opening reactions. These can convert cycloaliphatics or aromatic compounds to branched aliphatics. This would facilitate conversion of various components of diesel fuel to jet fuels of different types. Accomplishing this would require improved catalysts. The technology appears to be feasible, but there is no real economic incentive to accomplish it in the civilian market.

Changing the scale of a process to provide higher heat and mass transfer or to optimize yield is a developmental approach that appears to be predominantly a chemical engineering challenge. However, at the lower end of the size scale, it devolves into solving materials-related engineering problems because the size of components must become quite small. Here, large numbers of MEMS or mesoscale processors are used in parallel to improve yield, reduce emissions, provide more rapid heat and mass transport, enhance reaction rate, and improve safety. This microscale process intensification approach could result in revolutionary rather than evolutionary processing capabilities if the design and manufacturing difficulties associated with combining large numbers of very small processors and components into a single practical system can be overcome.

In summary, for logistic fuels most changes through 2020 will remain evolutionary. Potentially revolutionary advances in fuels will come from using micro/MEMS-based chemical approaches to processing logistic fuels. Other advances that relate to methods for on-board reforming of logistic fuels and for storing fuels like hydrogen are discussed in the section on fuel cells in Chapter 4.

FUEL CELLS

Description

Fuel cells are devices that directly convert chemical energy into electrical energy and, because they do not involve combustion, are not Carnot-limited. The principle of operation of a fuel cell is illustrated in the text of the Energy and Power Materials Panel report (Chapter 4). Much as in a battery, the components in a fuel cell that make direct electrochemical conversion possible are an ion-conducting electrolyte, a cathode, and an anode. Fuel is fed to the anode and an oxidant to the cathode. Oxidation of the fuel proceeds electrochemically via "half-cell" reactions that take place at each electrode. The rate at which these reactions occur, and thus the efficiency of the fuel cell, depends on the performance of the catalysts incorporated into the electrodes. Fuel cell efficiency is also directly related to the area-specific conductivity (conductivity/thickness) of the electrolyte. In the case of liquid electrolytes, the electrolyte is infused into some sort of matrix or sponge material, the surface properties of which are optimized to retain the corrosive electrolyte liquid within its structure.

Several different types of fuel cells have appeared over the past few decades; they are differentiated essentially by the type of electrolyte they employ (Table D-2), which in turn determines the temperature at which the fuel cell is operated. For efficiency purposes, higher temperature operation is preferred, but for portable power applications, lower temperature operation is preferred. High temperatures, as employed for molten carbonate and solid oxide fuel cells, generally ensure that reactions are rapid and a variety of hydrocarbon fuels can be used directly in the fuel cell. However, start-up times can be very long, and there is a limited number of (usually expensive) materials available for fabricating the fuel cell. Low

TABLE D-2 Fuel Cell Types and Selected Features

Type	Temperature °C	Fuel	Electrolyte	Mobile Ion	Recycled Species
Polymer electrolyte membrane (PEM)	70-110	H_2, CH_3OH	Sulfonated polymers (Nafion™)	$(H_2O)_nH^+$	H_2O
Alkali fuel cell (AFC)	100–250	H_2	Aqueous KOH	OH^-	H_2O
Phosphoric acid fuel cell (PAFC)	150-250	H_2	H_3PO_4	H^+	
Molten carbonate fuel cell (MCFC)	500-700	Hydrocarbons, CO	$(Na,K)_2CO_3$	CO_3^{2-}	CO_2
Solid oxide fuel cell (SOFC)	700-1000	Hydrocarbons, CO	$(Zr,Y)O^{2-}$	O^{2-}	

SOURCE: Hirschenhofer et al. (1998); Larmine and Andrews (2000); and Williams, M., "Overview of Fuel Cells," briefing presented to the Panel on Energy and Power Materials of the Committee on Materials Research for Defense After Next, National Research Council, Irvine, CA, October 10, 2001.

temperatures, as employed in proton exchange membrane fuel cells, are advantageous for intermittent power applications because start-up times are short, but at low temperatures chemical reactions are slow and fuel choices are essentially limited to hydrogen and possibly methanol.

While there are several technological hurdles to be surmounted before fuel cells become commonplace in either the civilian or military sectors, they offer several attractive features. In particular, (1) they are highly efficient, easily 45 percent for high-temperature fuel cells with even greater efficiencies in hybrid systems; (2) because of their operation on hydrocarbon fuels, they offer the potential for very high energy and power densities; (3) they are easily scaled, without significant loss in performance parameters, to a wide range of power demands; (4) they are nonpolluting (NO_x and SO_x emissions are ultralow); (5) they can be implemented in a distributed, modular fashion, resulting in a power generation system that is robust under combat conditions; and (6) they can be operated in a closed-cycle system that uses as input solar energy or fuel cell waste products to generate chemical fuels suitable for reuse in the fuel cell. This mode of operation is particularly advantageous for remote military operations, in which reliance on delivery of fuel implies a significant vulnerability.

Each of the fuel cell types listed in Table D-2 has been demonstrated in complete fuel cell systems, with AFC and PAFC being the most mature technologies and SOFC and PEMFC the newest. The choice of electrolyte places fundamental constraints on fuel cell system design. First, if a liquid electrolyte (alkali, phosphoric acid, or molten carbonate) is used, there must be some method of containing the electrolyte, which is corrosive to most materials, yet ensuring its access to fuel and oxidant gases. Second, if the ion transport relies on a "vehicle" mechanism—carbonate ions to shuttle oxygen (MCFC), hydronium ions to shuttle protons (PEM), or hydroxyl ions to shuttle oxygen (AFC)—then the fuel cell system inherently requires recycling of an inert gas—CO_2 in the first case or H_2O in the other two cases. Third, if we compare "true" proton conductors (not hydronium ion conductors) and oxygen ion conductors, both result in equally simple fuel cell systems, so long as the fuel is hydrogen, but for hydrocarbon fuels, SOFCs offer simpler systems because oxygen is supplied to the fuel and complete reaction to form CO_2 and H_2O is in principle possible. In the PEM system, protons are "stripped" from the fuel; to avoid carbon deposition, additional water must be fed to the anode to encourage reforming of the fuel (usually methanol) into CO_2 and H_2. Thus, oxygen ion-conducting electrolytes offer the potential for extremely simple fuel cell systems, although in practice, this has not been achieved.

Demonstrated fuel cell system efficiencies are in the range of 40 to 45 percent, and much higher efficiencies are expected when waste heat recovery schemes are used. Today's system costs range from ~\$4,500/kW for phosphoric acid and molten carbonate fuel cells to over \$10,000/kW for solid oxide, polymer electrolyte, and alkali fuel cells. In the Partnership for the Next Generation of Vehicles program (now abandoned), PEMFC costs were targeted to drop to \$50/kW by 2004. DOE analyses predict that costs of ~\$800/kW can be reached in solid oxide systems by 2005.[7]

Fuel Cell Research Efforts Required to Meet DoD Needs

This section expands on the breakthroughs needed in fuel cell research identified in the main body of the report.

Electrolytes and Ion Transport

A key limitation of fuel cell performance for all types of fuel cells is the resistance of the electrolyte, particularly in solid oxide fuel cells, for which high resistivity requires high-temperature operation, which in turn requires use of costly materials for the other fuel cell components. Research efforts to enhance the ionic conductivity of oxides and thus enable reductions in operating temperature (~600°C) are under way at many university and government laboratories. These efforts have yielded new materials like $(La,Sr)(Ga,Mg)O_3$ (LSGM), an oxide ion conductor, and doped $BaCeO_3$, a proton conductor, leading to optimization of known materials like doped CeO_2, an oxide ion conductor. So far, few studies of these materials as electrolytes in single-cell fuel cells have been carried out; presumably, their high conductivities will translate directly into increased fuel cell efficiencies, but little is known about the appropriate electrocatalysts for such systems. Given the broad range of oxide materials already known but unexamined as electrolytes, and those yet to be discovered, the potential for uncovering materials with enhanced ionic conductivity, excellent chemical stability, and good mechanical integrity is tremendous.

---

[7]Williams, M., "Overview of Fuel Cells," briefing presented to the Panel on Energy and Power of the Committee on Materials Research for Defense After Next, National Research Council, Irvine, Calif., October 10, 2001.

Polymer electrolyte membrane fuel cells rely on relatively facile transport of hydronium ions through hydrated regions of the sulfonated polymer. Their high conductivity allows for operation at close to ambient temperatures. The hydrated nature of the polymer electrolyte gives rise to some daunting challenges: (1) because the mobile species is hydronium, high humidification is required to maintain high conductivity, yet this must be balanced by the need to remove water from the cathode and prevent flooding of the electrocatalyst, (2) the maximum temperature of operation is ~100°C, (3) methanol diffuses from anode to cathode in a direct methanol PEM fuel cell, and (4) mechanical properties degrade due to "swelling" upon hydration. As with SOFCs, a broad-based research effort is under way to develop polymeric materials that do not suffer from these drawbacks. Indeed since 1995, the automotive industry has invested ~$2 billion worldwide in PEM fuel cells, with a significant portion of that investment directed at improving membranes. In many cases, however, success in, for example, reducing methanol crossover has come at the price of reduced conductivity. A radically different approach is to use anhydrous inorganic proton conductors based on acid salts (e.g., $CsHSO_4$) that truly transport protons rather than hydronium ions. While this approach has demonstrated success, limitations due to electrolyte solubility in water and degradation under fuel cell operating conditions are yet to be addressed (Haile et al., 2001).

In the case of molten carbonate, alkali, and phosphoric acid fuel cells, it is fair to conclude that further significant advances in the electrolyte material, particularly in ionic conductivity, are unlikely. Advances are more likely to come about through enhancements in the stability of the matrix material (through modifications to both the electrolyte and matrix compositions) and via processing routes that result in matrix structures with well-defined and reproducible porosities and surface areas. These advances are considered evolutionary rather than revolutionary.

Exploratory synthesis of new solid-state electrolyte materials, guided by computational materials science and enhanced by combinatorial approaches, thus constitutes an area ripe for producing revolutionary advances in fuel cells for future defense systems. Solid-state conductors of either oxide ions or protons with conductivities of ~10^{-3} S/cm at temperatures between 25 and 500°C and sufficient mechanical and chemical integrity would completely alter the fuel cell landscape. Direct methanol fuel cells (DMFCs) are a good example. Because they are relatively simple and have few auxiliary components, they are considered most suitable for single-soldier applications. In today's DMFCs, power densities are low because the methanol concentration in the fuel, a mixture of water and methanol, is limited to no more than ~3 percent and the electrolyte is relatively thick. Higher concentrations and thinner electrolytes lead to dangerous levels of methanol crossover (diffusion across the hydrated polymer), which can cause uncontrolled combustion. Zero methanol crossover electrolytes would reduce the amount of unnecessary water carried by the soldier by a factor of about four while increasing the power density of the fuel cell by at least a factor of two. Though civilian applications would also benefit from such electrolytes, the option of alternative fuel cell types in many of those applications (e.g., reformate-based PEMFCs) lowers civilian motivation to address this challenge. In sum, DoD investments in perovskite and fluorite oxide ion conductors, perovskite proton conductors, and anhydrous acid salts as pathways toward revolutionary new electrolyte materials are warranted.

Electrodes and Electrocatalysis

Critical to the function of a fuel cell are the electrodes and electrocatalysts, which catalyze the electrochemical reduction of the oxygen at the cathode and the oxidation of the fuel at the anode. The relatively poor performance of today's anode catalysts necessitates the use of hydrogen, or possibly methanol, as the fuel in lower temperature fuel cells (PEM, PAFC, and AFC). Indeed, in PEM fuel cells the hydrogen must be free of impurities in order to ensure proper catalyst performance (<50 ppm CO and <50 ppm H_2S).[8] The low operation temperatures of these fuel cells, along with the chemically aggressive nature of the electrolyte, further limit electrocatalyst choices to costly precious metals, except in alkali fuel cells. Inadequate anode catalyst performance is also what limits the use of diesel fuels in higher temperature (SOFC and MCFC) fuel cells, although clean alkanes (methane, ethane, propane, etc.) can be readily used. Poor cathode catalyst performance lowers fuel cell power densities, thereby making necessary larger and more costly systems, but generally does not affect the choice of fuel.

Breakthroughs in fuel cell electrocatalysts would revolutionize fuel cell technology, but it is not at all clear how such breakthroughs can be

[8]Bellows, R., "Fuel Processing Performance and Evaluation," paper presented at the NSF Fuel Cell Workshop, Arlington, Va., November 14-15, 2001.

achieved. Extensive research, from fundamental studies of H_2 adsorption on Pt single crystals to Edisonian "try-and-tinker" approaches to new catalyst alloys, is currently under way in laboratories worldwide. Some of these efforts have already led to important changes in our understanding. Two are highlighted here.

It has long been believed that the use of hydrocarbon fuels in solid oxide fuel cells requires an internal reforming reaction step, in which water and fuel react in the vicinity of the fuel cell anode to produce CO_2 and H_2, and that the hydrogen so generated is then consumed electrochemically to produce electricity. Groups at Northwestern University (Murray et al., 1999) and the University of Pennsylvania (Park et al., 2000) have independently shown that, in fact, water is not required for fuel cell operation. These results have radically changed the picture of anode reaction pathways and suggest that simpler fuel cell designs in which water recirculation is not necessary are on the horizon.

Perhaps even more significant have been reports from the Industrial Research Institute of Nagoya, Japan, indicating that fuel cell reactions can be so well controlled by the electrocatalysts that fuel and oxidant need no longer be separated (Hibino et al., 2000). The simplifications that might be afforded by such a "single chamber" fuel cell system are breathtaking. It is critical that the reliability of these results be verified by independent laboratories.

The importance of electrocatalysts, and the absence of a clear pathway for success, dictate that the military explore a broad range of solutions to address the challenge of improving fuel cell catalysts, particularly diesel fuel anode catalysts, which are of lesser interest to the civilian sector. Many tools developed in the last few years can and should be effectively leveraged by DoD as part of this effort: (1) combinatorial chemistry methods for quickly finding and optimizing new alloy compositions for catalysts, (2) computational materials chemistry for guiding exploratory research in identifying new catalysts, and (3) new synthetic methodologies that allow for architectural control of fuel cell electrodes at the nanoscale, potentially enhancing reaction kinetics by dramatically increasing electrode surface area and restricting reactions to confined regions. The fuel cell community has yet to take advantage of advances in these areas, and it is at the interface between these fields and traditional electrochemistry that the most significant breakthroughs can be anticipated.

Fuels and Fuel Cell System Design

It is evident from the previous discussion that the choice of fuel for fuel cell systems is a fundamental question. While higher temperature fuel cells can operate directly on hydrocarbon fuels (typically using water to achieve internal reforming), lower temperature fuel cells operate on hydrogen or possibly methanol. Generating, storing, and transporting hydrogen is a significant hurdle. Hydrogen is typically produced from methane (CH_4), which is reacted with water (reforming) or oxygen (partial oxidation) at high temperature to produce a mixture of CO_2, CO, H_2, and residual CH_4. The CO in this mixture is then reacted with H_2O at lower temperatures to yield CO_2 and H_2 via the "water gas shift." Further reductions in CO concentration can be achieved by preferential oxidation (reaction with oxygen again).

Where the starting material is some other fuel, sulfur removal is also an important step. Removal of trace levels of CO is of paramount importance for PEM fuel cells, whereas trace levels of sulfur are problematic for all fuel cell types. These impurities adsorb onto the surface of the anode catalyst, reducing fuel cell voltage and thus power output and efficiency.

Today's fuel reformers are extremely efficient (~85 percent) because of tight thermal integration and effective catalysts but do not produce H_2 pure enough for long-term PEM fuel cell operation. Methane reformers are the most mature technology; gasoline reformers are under rapid development in the civilian automotive sector. For DoD applications, reformers that process diesel fuels for lower temperature fuel cells or high-temperature fuel cells that use diesel fuels are essential. Avoiding dependence of the military on imported fossil fuel sources may require effective strategies for coal gasification and subsequent hydrogen generation, as well as effective methodologies for sequestering CO_2, the natural byproduct of any energy production involving coal.

In recent years, there have emerged new approaches for generation of clean hydrogen from a variety of fuel sources that warrant further DoD investment. These rely on membrane reactors in which the reactants (for example, CO and H_2O in a water gas shift reactor) are fed into one side of a membrane that selectively transports one species (hydrogen in this example); the reaction products are then extracted from the other side of the membrane (Figure D-5). The effect of the water gas shift reaction is to both ensure that pure hydrogen is generated (no other species can transverse the selective membrane) and enable hydrogen production rates

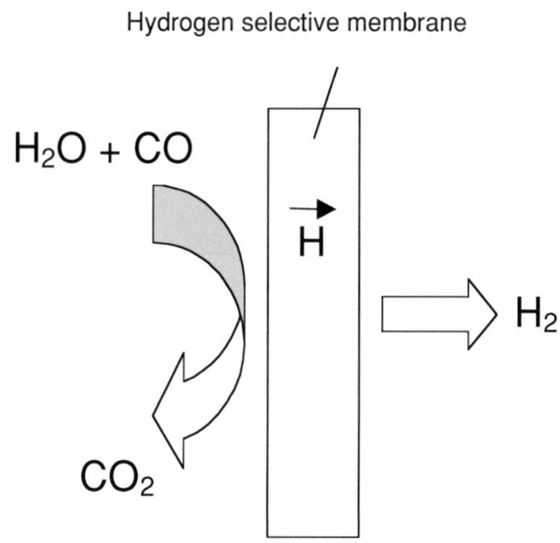

FIGURE D-5

Schematic of a membrane reactor, using the water gas shift reaction as an example. The overall reaction is $H_2O + CO \rightarrow H_2 + CO_2$. The right-hand side of this reaction is favored by rapid removal of H_2 from the reaction zone.

beyond the thermodynamic limit, because the hydrogen concentration in the reaction regime is kept very low. Successful operation of membrane reactors requires materials with high hydrogen fluxes, good catalytic properties, and excellent chemical and mechanical robustness. Research in this area is in its infancy, with most of it directed at palladium-based membranes and some limited work on ceramic membranes. Given the potential of such reactors to solve the daunting fuel problem for fuel cells, additional efforts in this direction are needed.

An entirely different strategy for providing hydrogen for fuel cells is the use of nonconventional fuels like ammonia, hydrazine, or sodium borohydrate. As with hydrogen, another energy source must be used to synthesize these fuels, but they offer the advantage of being processable in a closed cycle, so that reliance on fossil fuels is minimized. Particularly attractive is sodium borohydrate, which readily reacts with water to release hydrogen and the byproduct, sodium borate, can be regenerated using independent energy inputs.

INORGANIC THERMOELECTRIC MATERIALS

Thermoelectric (TE) materials have applications like thermal management, man-portable cooling, and recovery and conversion of waste heat into useful electrical power, all of which are of importance to DoD. Cooling devices made from these materials contain no moving parts or environmentally unacceptable working fluids like chlorofluorocarbons. Such devices can have high reliability, low weight and volume, and low acoustic signature while allowing for precise temperature control.

TE modules are based on materials with a unique combination of properties, including high thermopower but also high electrical conductivity combined with low thermal conductivity. This combination of properties is critical to high device efficiency but is difficult to achieve in a single material, because these properties are often quite interdependent, with (for example) high electrical conductivity often connoting high thermal (phonon) conductivity as well. This situation is both a great challenge and an exciting opportunity for solid-state chemists, theorists, and materials scientists to devise creative approaches to optimize all the desired properties.

The figure of merit for a TE material is defined as $ZT = (S^2\sigma/k)T$, where S is the thermopower (or Seebeck coefficient), σ the electrical conductivity, κ the thermal conductivity, and T the temperature. When a temperature gradient is applied to a thermoelectric material, the mobile charge carriers at the hot end preferentially drift to the cold end, inducing a voltage; this is called the Seebeck effect. The sign of the Seebeck coefficient depends on the sign of the charge carriers. Consequently, a pair of n-type and p-type "legs" can be placed in parallel with a temperature gradient but in series with respect to electron flow and the voltage generated across each leg will be additive; power is delivered to a load placed in the circuit from both legs. Useful TE materials must not only exhibit high ZT over a wide temperature range but must also be compatible with both donor and acceptor doping. Although the two legs could be made of unrelated n- and p-type compounds, complications with material compatibility and device design can arise if their properties (coefficient of thermal expansion, thermal conductivity, electrical conductivity, etc.) differ significantly.

While TE devices are highly desirable, they are not widely available due to their low efficiencies, a fact directly attributable to the low values of ZT in known materials. Even the best TE materials exhibit bulk values of ZT that are only on the order of 1, a milestone that was achieved nearly 40 years ago (Vining, 2001). For viable TE devices for DoD, values of ZT up to 4 are necessary. Such targets, while they have remained elusive, violate no laws of thermodynamics. Indeed, there is no fundamental limit to the possible value of ZT, so the slow progress of recent years calls for new approaches to TE materials.

The best materials are based on complex semiconductors with optimized carrier concentration, high effective mass carriers, high carrier mobility, narrow bandgap, and low lattice thermal conductivity (Nolas et al., 2001). Examples are Bi_2Te_3, PbTe, $Si_{0.8}Ge_{0.2}$, and Zn_4Sb_3. Unlike many technological solid-state electronic materials, in which simplicity in composition and structure and extreme crystalline order and purity are sought, TE materials are more likely to have complex composition, elaborate crystal structure, and even disorder, all features that may lower lattice thermal conductivity without sacrificing electrical conductivity and thermopower. Moreover, a single material is unlikely to exhibit high ZT over the entire temperature range of interest. Therefore, a broad-based and sustained exploratory synthetic effort that encompasses many classes of materials and structures (chalcogenides, pnictides, intermetallics, oxides, etc.) and even leads to entirely new families of materials must be pursued if the challenge of high ZT is to be met. Here solid-state chemistry is of central importance.

Nanoscience may also have a role to play in next-generation thermoelectrics. Certain theoretical considerations predict that quantum confinement, which can increase the density of states at the Fermi level, may be another path to systems with high ZT. Quantum confinement using multilayered structures can also provide between-layers interfaces that selectively scatter or confine phonons (lattice vibrations) to reduce thermal conductivity. Thin-film TE materials and devices thus offer exciting possibilities. Preliminary results for Bi_2Te_3/Sb_2Te_3 superlattices are very promising, with a reported effective ZT of 2.4 (at room temperature) (Venkatasubramanian et al., 2001); efforts must now be directed toward implementing such structures in functional devices to experimentally verify this concept.

Any exploratory synthetic effort toward new TE materials must be complemented with a strong computational/theoretical effort, particularly in electronic band structure calculations and interpretation. Several parameters are critical in establishing TE behavior: actual bandgap, the shape and width of the bands near the Fermi level, the carrier effective masses and mobilities; and the degeneracy of band extrema (i.e., the number of valleys in conduction band or peaks in the valence band). Of the three properties that define Z, thermopower is least understood. It cannot be predicted, nor is it known how to increase the thermopower of a material without depressing electronic conductivity. In enhancing understanding, solid-state theory can provide invaluable guidance in selecting materials for consideration. The synthetic effort must also be matched by breakthroughs in approaches and methods of screening large numbers of candidate materials reliably and quickly. This necessarily brings together the fields of solid-state chemistry, solid-state physics, materials science, and electrical engineering. Close collaboration between scientists from these areas is of paramount importance.

INORGANIC PHOTOVOLTAICS FOR ENERGY CONVERSION

Organic photovoltaic (PV) materials are discussed in Chapter 6. This section briefly addresses materials and processes for devices made from inorganic materials that are useful in converting light into electricity. These devices offer the potential for harvesting ambient solar energy and are of interest because the input energy is free and the process is nonpolluting. If sufficiently high efficiencies could be obtained, these devices would be revolutionary for DoD in simplifying logistics and modifying energy resupply requirements. In theory, solar radiation could supply electricity and also make it possible to produce hydrogen from available water to fuel hydrogen fuel cells. Even at lower efficiencies, these devices could provide some portion of the energy a soldier needs in combat.

PV devices convert light directly into electricity with no net chemical change left behind. They have been based on such materials as doped forms of single-crystal, polycrystalline, or amorphous silicon; compound semiconductors have been based on III-V materials. An alternate type is a photosynthetic cell with two redox reactions, one involving reaction with electrons at the counterelectrode and one with holes at the semiconductor electrode. This type of cell, also known as a photoelectrochemical (PEC) cell, can be used to generate electricity from light or can convert light into chemical potential energy via photoelectrolysis. While TiO_2 has been used successfully as a semiconductor in PEC cells, it absorbs primarily in the UV portion of the spectrum and has low conversion efficiency. Use of the cubic perovskite $SrTiO_3$ raises conduction band edge energy and allows spontaneous photoelectrolysis of water, which has been one objective of photoelectrochemists for years. However, the material still functions in the UV and efficiency remains too low for commercial use. Attempts to produce other oxides that are environmentally robust and that absorb preferentially in the visible range have been largely unsuccessful. Separation of the functions of optical absorption and charge generation has led to improved devices. This has been achieved by combining oxide semiconductors with ruthenium-based sensitizing dyes that absorb in the visible and facilitates injection of charge carriers into a stable (wide bandgap) semiconductor (Hamnett and Christensen, 2000).

Nanomaterials offer potential dramatic property improvements, here in terms of devices with higher efficiencies and enhanced environmental stability. Examples are dye-sensitized mesoporous film oxides or chalcogenides, which are composed of nanometer-sized crystals, are dye-sensitized, and have their porosity filled with a second medium that can be solid or liquid and that involves a semiconductor or a conducting electrolyte. The extremely large contact area when combined with dye sensitization can lead to much enhanced (1,000 times) capture of incident light and greatly increased photocurrents. Similarly, solid heterojunctions having large contact area can be grown from electron and hole-carrying materials in the form of interpenetrating networks and have demonstrated some promise (Grätzel, 2001). Future effort in this area should examine nanoscale structures like semiconductor quantum dots for efficient photon collection and rapid charge conduction, with a goal of increasing device efficiency while reducing manufacturing cost. This cost has slowed the adoption of PVs in the past and is one reason (together with low weight and ability to employ flexible substrates) that effort has been directed toward organic PVs. If PV approaches are to become feasible for DoD or the commercial arena, a low-cost means of making PV cells in large quantities is essential. Today's laboratory-scale techniques for generating promising material configurations must give way to truly manufacturable approaches.

ELECTRIC POWER GENERATION FOR FUTURE DOD SYSTEMS: EFFECTS OF ELECTRIC PROPULSION IN SHIP DESIGN

Platform propulsion and support using more electric systems is of importance to the Army, Navy, and Air Force. The consumption of electric power by elements of DoD will likely increase significantly over the next 25 years. More correctly, the *ability* to consume power will rise; with better efficiency, the actual increase in consumption can be mitigated, though several factors will push the overall electrical load higher. Using a U.S. Navy ship as an example, this section (1) summarizes why electric power use may increase; (2) shows how integrated electric propulsion might positively affect ship design; (3) identifies specific DoD technology needs;

and (4) briefly points out materials issues. The goal of this appendix is not to identify all likely materials, because the panel did not have the resources to do so. Rather, it was to give the interested reader a brief view of the potential effects of changes in electric propulsion on ship design.

Increased DoD Needs for Electrical Power Generation

Some key reasons that use of electric power is likely to increase in DoD systems are the following:

- *The force as a whole will have fewer uniformed personnel.* Eliminating sailor positions will decrease the direct and indirect personnel costs of defending America. However, the tasks to which those people were assigned may not be eliminated, so more automation, robotics, and power must replace them.
- *Components of the DoD will be greener.* While a major tenet of green philosophy is conservation, removing pollution from the environment is not without an energy cost. Preprocessing of fuels and post-processing of exhaust will push demand. Reducing a ship's waste stream or increasing shipboard air quality, for example, will require more electrical power.
- *Information technology (IT) will consume an increasingly large amount of power.* For example, aboard-ship IT consists of not only the ship's computer network, but also the transmitters and receivers to move information on and off the vessel. As signals become more complex, the power needed to generate and process signals must increase.

DoD is moving in the direction of more-electric systems, including electric propulsion. In January 2000 the Secretary of the Navy announced that the next generation of warships would be propelled by an electric drive system.

Advantages of Electric Propulsion in Ship Design

Compared to the standard mechanical drive system that uses reduction gears, there are many advantages to an integrated electric propulsion system:

- *Increased power flexibility* allows significant amounts of power to be directed either to the ship's service power system, to the ship's propulsion system, or to weaponry, depending on the situation.
- *Increased design flexibility* eliminates the reduction gear and the drive train, opening a completely new design space for future ships and allowing large, heavy power-generating equipment to be located anywhere in the ship.
- *Reduced total ownership cost* means that power generators can operate at a more efficient speed almost continuously, thus significantly increasing efficiency while reducing fuel cost and engine room staffing.
- *Leveraged commercial electronic and electromagnetic technology* replaces reliance on a mechanical drive manufacturing base that no longer produces commercial vessels, with reliance on equipment (e.g., insulated gate bipolar transistors) already available from the commercial power industry.
- *Commonality* allows major pieces of electric propulsion equipment to be used in virtually every warship in the U.S. Navy, facilitating modular spare parts usable across platforms.
- *Reduced environmental emissions* result from integrating the ship's service and propulsion power systems. Remaining prime movers can operate at high efficiency, reducing exhaust emissions.
- *Enhanced modularity* provides parallel design processes and facilitates modernization and repair on vessels while significantly reducing manufacturing cost.
- *Reduced overall maintenance requirements* allow prime movers to operate at their peak performance range at relatively constant power levels for long periods.
- *Increased payload capacity* reduces fuel load out, allowing the platform more weapons or ones with greater range, or keeping them longer on station time.
- *Increased automation* means that automatic or autonomic systems carry the burden of operating the system, reducing the number of people in the loop.
- *Flexibility/upgradability* allows DoD to replace existing combat systems with new ones that use significantly more power without significantly affecting maximum vessel speed.

DoD-Specific Technology Needs

An array of issues must be addressed before electric propulsion systems become a reality. Many are unique to DoD and will require new approaches to system design and architecture, advanced modeling of electromagnetic characteristics, and, equally important, advances in new materials and processes for making them. Some of the major considerations from the DoD perspective include:

- *Power density.* The real estate aboard a military vessel is in high demand, yet DoD is challenged to insert more payload into smaller and cheaper vessels. Because space requirements in today's commercial market are not as restricted the military cannot leverage on industry.
- *High power levels.* One way the marine industry is attempting to mitigate the power density problem is by raising the power levels associated with the propulsion system, but this raises a myriad of issues not previously experienced by marine propulsion engineers. The potential power required of DoD weapon systems only exacerbates the problem. Thermal management becomes an even more critical design criterion and safety of operators and maintenance workers demands more attention.
- *Electromagnetic interference.* As automation pervades every area of combat, electromagnetic interference soon becomes critical. Power conditioning equipment is susceptible to the electrical noise generated by the high current and high voltage loads and sources that surround it. Electromagnetic characteristics can also contribute to the ship signature. The commercial industry does not support research in this area, but it is critical to DoD.
- *Shock and impact.* This area receives no attention from industry, but it clearly must be addressed for DoD.

Related Materials Issues

A key near-term concern is high-frequency switching and the associated thermal management problems in a power-dense environment. While silicon carbide switching technology is now under development, by 2020 it may be necessary to have available additional options for high-temperature and efficient power-switching and conditioning components. Power electronics were discussed in Chapter 5. Many of the materials technologies already described may help solve these large-scale power problems. For example, increased use of DC power components could eliminate the need for power conditioning and simplify its distribution. DC promotes the use of components whose raw electrical output is DC, such as fuel cells, and DC has attendant benefits in terms of signature reduction, survivability, and damage control.

The need to carry large amounts of power at low loss while minimizing the volume of power carriers argues for advanced high-temperature superconductors with high critical currents. Similarly, the need for minimizing volume aboard military platforms argues for high-temperature superconducting (HTS) motors, which will have a profound impact on ship architecture, performance, and staffing requirements. These motors should have major advantages over conventional induction motors: It is projected that they will be only 20 percent of the size and 33 percent of the weight of conventional motors. If built in the nearer term, such motors would likely use HTS ceramic superconductors in a metal matrix operating at 77 K (liquid nitrogen). New MgB_2 superconductors show considerable promise but at present have a T_c of only 39 K, and would probably run in liquid hydrogen coolant gas. While liquid hydrogen might be acceptable in a civilian application, there are serious concerns about producing and storing significant amounts of it on a warship, where hydrogen's high explosivity would be a real issue in combat. Nevertheless, as MgB_2 and other alloys appear, the T_c may increase to the point where inert liquid gases could provide cooling.

Other materials developments useful for all-electric platforms are advanced dielectric materials for capacitors and high-strength composites for flywheels and compulsators; specific materials developments identified for these components are described elsewhere in the report. The panel did not have the resources to address in depth all the materials issues in this area, but recommends that this be done once a preliminary design for an all-electric platform is completed and device operating parameters are identified. This may result in identifying new materials needs in such categories as high-temperature (1000°F) low-loss electrical insulation; high-performance heat sinks; or EMI shielding materials.

Integrated platform electrical power and propulsion systems will also require improved modeling and analytic tools and advanced algorithms to model effects and interactions among systems and material characteristics in this highly complex environment. Detailed system modeling tools must reflect integrated power-system control and transient response as well as modeling local components. It is vital to understand how electromagnetic parameters scale differently from either hydrodynamic characteristics or thermal parameters; scaling the electric propulsion system is a challenging task, but there is a shortage of empirical data to support this type of modeling, and the database must be more robust if development costs for full-scale electric propulsion systems are to be contained.

PROJECTED LIGHTWEIGHT ARMOR AREAL DENSITY GOALS

Table D-3 presents projected areal density (AD) goals for lightweight body armor and transparent armor in the period 2000-2030. The main body of this report contains additional information.

DOD RELIANCE ON ENERGY SOURCES

National energy dependence, particularly reliance on energy imports, has been the subject of considerable policy discussion for many years and national energy source and consumption trends continue to be discussed, debated, and addressed at the highest levels in this country because of their implications for national security. While this is not strictly a materials issue, the committee agreed that a brief discussion of total DoD energy usage, dependence on various sources, and energy dependence through 2020 was desirable. This section examines the major energy sources: fossil fuels and nuclear energy.

TABLE D-3 Goals for Future Armor Areal Density

		Areal Density (lb/ft^2)		
Armor Type	Threat	Present	Projected, 2010–2015	Projected, 2020–2030
Fabric torso protection	Fragments	1.38	1	0.75
Rigid fabric helmet	Fragment	2.12	1.7	1.5
Rigid fabric/composite torso	0.30-caliber ball	6.1	3.5-4.5	<3.5
Torso	0.30 AP	6.5-7.5	<5	<3.5
Transparent	0.30 AP	23-33 (glass/polymer)	<10.5 (ceramic/polymer)	<8 (ceramic/polymer)

SOURCE: S. Wax, DARPA, personal communication; J. Ward and P. Cunniff, Soldier Systems Center, Natick, personal communication; Cunniff (1999a,b); and DARPA (1998).

Dependence on Fossil Fuel

In FY97 total U.S. energy consumption from all sources was 94.21 quads (1 quad = 1×10^{15} British thermal units [BTUs]). For that year, total federal government usage of energy was 1.53 quads. DoD used 1.13 quads in fossil fuel energy, approximately 1.2 percent of the U.S. total (DOE, 1999)—down from 1995, when DoD energy usage was 1.4 percent (1.19 quads) of the U.S. total of 87.3 quads (DOE, 1997).

Of the DoD total for FY97, 58 percent (approx. 0.65 quads) was used for actual support of military operations and training, with the remainder used for buildings and nontactical vehicles. Major energy sources were traditional fuels for military platforms (e.g., jet fuel, diesel) and natural gas and electric power supplied by a variety of land-based systems for fixed installations. No matter how it is calculated, it is clear that DoD energy usage to support the warfighter and military platforms is a very small percentage of total U.S. energy consumption. In addition, during a major crisis national security needs will be given priority, and DoD should have access to whatever energy is needed to assure continuity of its operations. While it is therefore likely that sufficient energy will be available to the DoD in future crises, a question remains about the form in which that energy would be supplied. In particular, will DoD energy consumption *patterns* change between now and 2020 and, if so, how?

First, 42 percent of DoD's energy consumption today is supplied to buildings, other fixed assets, and nontactical vehicles. Most of these installations are located within this country, and the energy sources supplying their needs are likely to be those supplying energy to other U.S. infrastructure. The current strong push within the government to reduce energy consumption per square foot by using high-efficiency energy systems and smart technology is expected to continue. At the same time, there is a push to reduce the number of installations by consolidation; this should reduce energy consumption for DoD-owned real property and buildings. Increased efficiency in use, transmission, storage, and production, combined with reduced demand and more efficient systems, offer considerable potential for reduction in this DoD sector.

The other 58 percent of DoD energy consumption is used to power military platforms for operations and for training (the latter being a major contributor to energy consumption, as it is present during peacetime as well as war). There is a continuing push here, too, for enhanced fuel efficiency for many reasons. A recent report of the Defense Science Board

(2001) examined DoD fuel reliance and found that greater fuel efficiency could lead to many benefits, among them enhanced platform stealth, reduced logistics tail, reduced vulnerability of supply lines, and an ability to build up forces faster. As an example, one goal of the Army transformation is to field a future combat system that is lighter weight, more mobile, and more fuel-efficient, because fuel today accounts for 70 percent of the Army tonnage that must be shipped.

There are several ways to reduce DoD fossil energy dependence. More use of alternative energy sources is one. Nuclear and solar power both have potential for certain DoD applications, and geothermal may have some naval applications. Wind and hydropower are unlikely to provide primary power for major platforms. Novel materials can affect many of these modes and may reduce DoD energy consumption through more efficient propulsion technologies; lightweight, high-specific-stiffness and strength structures; vehicle armor approaches that do not rely solely on areal density for protection but use lighter materials and novel protection schemes to reduce weight; higher-energy-density batteries, and fuel cells capable of direct electrochemical conversion; improved fuels; and smart electronics that use energy more effectively. Higher efficiency energy-harvesting materials and devices (e.g., PV conversion) might also augment available energy in specific circumstances.

The full panel report (Chapter 4) discussed the reliance on logistic fuels. It was concluded that both jet fuel and diesel would still be required by 2020, augmented perhaps by methanol, hydrogen (reformed from logistic fuels or otherwise manufactured), and possibly other hydrocarbons. That conclusion is discussed in the fuel cell section of the chapter.

In summary, DoD fossil fuel energy consumption currently represents only a small fraction of total U.S. energy consumption. Because of the systems and infrastructure already in place, it is likely that major energy sources currently used will still be of primary importance in 2020. In particular, fossil fuels will remain crucial to the majority of land and aerospace propulsion applications. While DoD energy sources are likely to reflect those used by society at large, the gradual introduction of new materials, systems, and capabilities and more efficient technology will improve energy usage. In the case of fuel cells, there is likely to be a gradual shift toward alternative fuels by 2020. On the whole, however, it is likely that current fuels (JP-8, diesel, and nuclear) will still make up the bulk of energy sources used by DoD systems in 2020.

Dependence on Nuclear Energy

Within the DoD, the Navy has been the primary user of nuclear power. The decades of nuclear power effectiveness that emerged from the early days of Adm. Rickover's Nautilus program have been impressive. In 1986 the U.S. Navy was operating or building 6 aircraft carriers, 9 cruisers, 39 ballistic missile submarines, and 97 attack submarines, all with nuclear propulsion plants, without any major incidences or measurable release of contamination.

Removing warship reliance on fossil fuel provides a significant military advantage. America's military commitments will continue to be worldwide. The ability of our nation's foremost projectors of power, the nuclear aircraft carriers, to transit to any location in the world at sustained flank speed is currently hindered only by the fact that the support ships that must accompany them are conventionally powered. America's nuclear-powered vessels are able to remain on station indefinitely, while the fossil-fueled vessels require significant infrastructure for fuel support.

Over the past 30 years, much of the controversy over nuclear power has been focused on civilian nuclear facilities. High-visibility accidents that released radioactivity at Three Mile Island, Chernobyl, and recently in Japan have heightened public concern about the safety, environmental impact, and total cost associated with harnessing nuclear energy. This directly affects DoD strategy for further acquisition of nuclear-powered assets. Indeed, only the unique requirements of aircraft carriers and submarines justify the cost of incorporating nuclear propulsion plants today. However, the current revolution in warship propulsion system design combined with the evolving all-electric warship offer an opportunity for DoD to reevaluate its use of nuclear power as an energy source. Consider the following:

- Today the world's energy comes from petroleum (~40 percent), coal (~25 percent), natural gas (~25 percent), hydroelectric (~5 percent), and nuclear (~5 percent).
- Well over 95 percent of the mobile power generated by DoD platforms (tanks, boats, ships, trucks/autos, aircraft) use petroleum.
- DoD mobile platforms rely predominantly on fuel oil from sources that the United States does not control.
- Because even the most efficient fossil fuel engines consume their own weight in fuel within half a day, fuel storage determines overall

systems weight, and fuel supply dominates logistics concerns.

- Over nearly 50 years, nuclear accidents have been few and the DoD track record on nuclear safety has been nearly flawless.
- Investment in nuclear power R&D is low compared with investment in oil, gas, wind, and photovoltaics, yet it has been estimated that, with sufficient investment, nuclear generation of electricity can be cheaper than generation by a combined-cycle gas plant.
- DoD's strategic vision of an all-electric warship/combat vehicle highlights advantages of nuclear generation for (1) warfighting effectiveness, (2) reduced total ownership costs, and (3) increased survivability.

Nuclear electric generators may significantly increase specific power output. A new generation of high-power-density nuclear power plants is possible but would require substantial R&D investment. Feasibility studies of thermonuclear-PV direct energy conversion being conducted at the Knoll's Atomic Power Laboratory show the clear potential of this technology.

As noted elsewhere in this report, fuel cells are considered a desirable energy source for future defense systems. The continuously renewable hydrogen used by fuel cells of the future could come from nuclear power plants (see the section above on fuel cells). Near-term refinement and deployment of fuel cell power plants, using the existing fossil-fuel delivery infrastructure would make possible long-term use of fuel cell power plants by all facets of DoD. Fusion reactors continue to show great promise. These reactors do not have the dangerous radioactive byproducts of the fission reactor, but they are technologically much more challenging.

Dependence on liquid fossil fuels began in the last half of the 19th century and throughout the 20th mankind has relied on them heavily. However, their price stability and availability cannot be guaranteed in the 21st century. Diversification of DoD energy sources to use more nuclear power is strategically prudent as well as technically and economically justified. Advances in the materials used in electromagnetic machinery, such as high-field permanent magnets and high-temperature superconductors, present opportunities to increase electrical power density and efficiency. Some of the potential for materials developments in these areas is discussed elsewhere in this report. The Energy and Power Panel did not have the resources to fully investigate materials challenges in this area but believes that key DoD platforms are likely to require nuclear energy in 2020 and beyond, and the rationale for increased attention to nuclear power systems and associated materials for DoD platforms is sound.

Technologies linked with advances in nuclear power that require further exploration include:

- Material technologies for motors and generators, such as superconducting magnets, cryogenic coolers, current collectors, high-field permanent magnets, liquid cooling, and active noise control;
- Materials with high thermal conductivities that are radioactively inert;
- Fusion reactor materials (likely to stretch beyond 2020);
- Materials and concepts for acceptable nuclear waste disposal and storage;
- Materials for very low harmonic motor controllers;
- Materials and designs for high-power-density and high-performance solid-state inverters and converters; and
- Electrical substitutes for systems and components that now rely on fluid transport for energy and actuation.

REFERENCES

Alchavan, J.A. 1998. The Chemistry of Explosives. Cambridge, UK: Royal Society of Chemistry Information Services.

Clelland, I., R. Price, and J. Sarjeant. 2000. Advances in capacitor technology for modern power electronics. Pp. 145-148 in Proceeding of the 24th International Power Modulator Symposium, June, Norfolk, VA. Piscataway, NJ: IEEE.

Cooper, P.W., and S.R. Kurowski. 1996. Introduction to the Technology of Explosives. New York: VCH Publishers.

Cunniff, P. November, 1999a. Dimensionless parameters for optimization of textile-based body armor systems. Proceedings of the 18th International Symposium on Ballistics, San Antonio, TX. Lancaster, PA: Technomic.

Cunniff P. November, 1999b, Assessment of small arms (ball round) body armor systems. Proceedings of the 18th International Symposium on Ballistics, San Antonio, TX. Lancaster, PA: Technomic.

DARPA (Defense Advanced Research Projects Agency). 1998. DARPA/ARO/ARL Transparent Armor Materials Workshop Proceedings, November 16-17, 1998, Annapolis, MD.

Defense Science Board, Task Force on Improving Fuel Efficiency of Weapons Platforms. January 2001. More capable warfighting through reduced fuel burden. Washington, DC: Office of the Under Secretary of Defense for Acquisition, Technology and Logistics. Available online at <www.acq.osd.mil/dsb/fuel.pdf>. Accessed April 21, 2002.

DOE (Department of Energy). 1997. Annual Report to Congress on Federal Government Energy Management and Conservation Programs for Fiscal Year 1995, U.S. Department of Energy.

DOE. 1999. Annual report to Congress on Federal Government Energy Management and Conservation Programs for Fiscal Year 1997, U.S. Department of Energy.

Federoff, B.T. 1960. Encyclopedia of Explosives and Related Items, B.T. Federoff, ed. Dover, NJ: Picatinny Arsenal.

Grätzel, M. 2001. Photoelectrochemical cells. Nature 414(6861):338-344.

Haile, S.M., D.A. Boysen, C.R.I. Chisholm, and R.B. Merle. 2001. Solid acids as fuel cell electrolytes. Nature 410(6831):910.

Hamnett, A., and P. Christensen. 2000. Energy conversion. P. 432 in The New Chemistry, N. Hall, ed. Cambridge, U.K.: Cambridge University Press.

Hibino, T., A. Hashimoto, T. Inoue, J.-I. Tokuna, S.-I. Yoshida, and M. Sano. 2000. A low-operating-temperature solid oxide fuel cell in hydrocarbon-air mixtures. Science 288(5473):2031-2033.

Hirschenhofer, J.H., D.B. Stauffer, R.R. Engelman, and M.G. Klett. 1998. Fuel Cell Handbook, Fourth Edition. Parsons Corp. Available online at <http://www.fuelcells.org/library/FCHandbook.pdf>. Accessed March 3, 2002.

Kornmann, X., L.A. Berglund, J. Sterte, and E.P. Giannelis. 1998. Nanocomposites based on montmorillonite and unsaturated polyester. Polymer Engineering and Science 38(8):1351-1358.

Larmine, J., and D. Andrews. 2000. Fuel Cell Systems Explained. Chichester: John Wiley & Sons.

Messersmith, P.B., and E.P. Giannelis. 1995. Synthesis and barrier properties of poly(e-caprolactone)-layered silicate nanocomposites. Journal of Polymer Science: Part A: Polymer Chemistry 33:1047-1057.

Murray, E.P., T. Tsai, and S.A. Barnett. 1999. A direct-methane fuel cell with a ceria-based anode. Nature 400(6745):649-651.

Nolas, G.S., J. Sharp, and H.J. Goldsmid. 2001. Thermoelectrics: Basic Principles and New Materials Developments. Berlin: Springer-Verlag.

Park, S.D., J.M. Vohs, and R. Gorte. 2000. Direct oxidation of hydrocarbons in a solid-oxide fuel cell. Nature 404(6775):265-267.

Venkatasubramanian, R., E. Siivola, T. Colpitts, and B. O'Quinn. 2001. Thin-film thermoelectric devices with high room-temperature figures of merit. Nature 413(6856):597-602.

Vining, C.B. 2001. Semiconductors are cool. Nature 413(6856):577-578.

APPENDIX E

Functional and Organic Hybrid Materials

ORGANIC BUILDING BLOCKS FOR CHROMOPHORE STRUCTURE AND PHOTOREFRACTIVE COMPOSITE MATERIALS

Common electron donor, π electron bridge, and electron acceptor functional groups are shown in Figure E-1, and a very small sample of typical electro-optic (EO) chromophore structures and their mb values are given in Figure E-2. Figure E-3 shows polymers and various molecular components used in preparing photorefractive composite materials.

SCHEMES FOR PHOTOREFRACTIVE POLYMERS AND ORGANIC MATERIALS

Photorefractive (PR) polymers and organic materials can be produced through several different approaches, as discussed in Chapter 6. Examples of two current approaches are shown in the next two figures. Figure E-4 shows several functional polymers reported to exhibit PR effects including functional polyurethanes, functional conjugated polymers, functional polyimides, and conjugated polymers containing transition metal complexes. Figure E-5 shows several amorphous molecular materials exhibiting high PR performances.

USING MOLECULES AS SWITCHES

Examples of simple molecular switches are shown in the next two figures. The photochromic molecule in Figure E-6 can undergo a photo-induced intramolecular change in its molecular

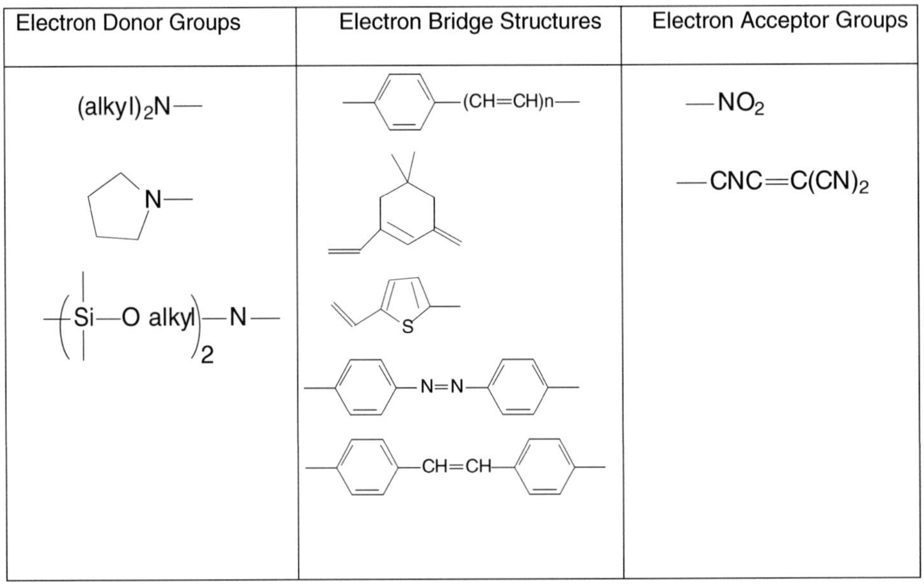

FIGURE E-1

Electro-optic chromophore building blocks.

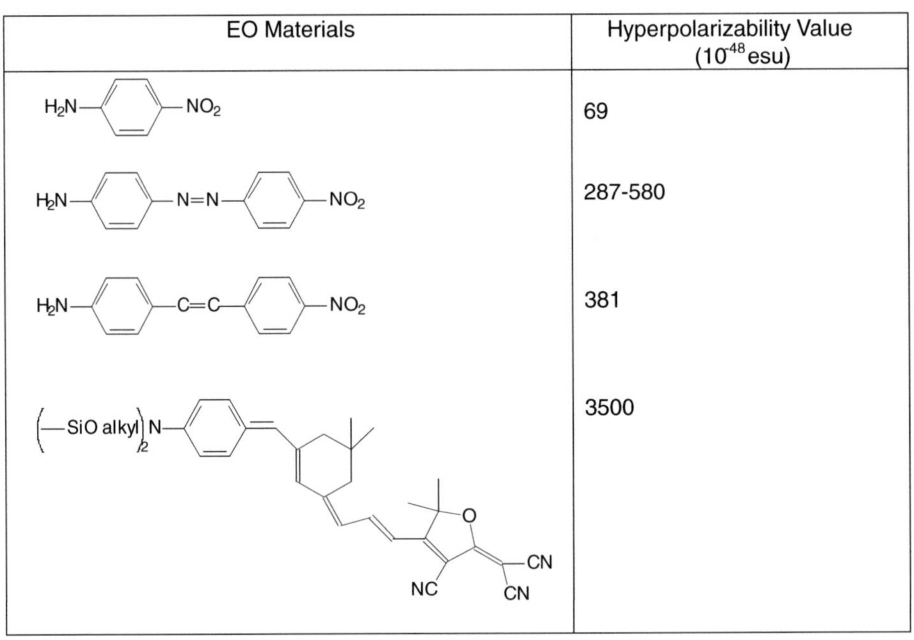

FIGURE E-2

Typical electro-optic chromophore structures and their first molecular hyperpolarizability values.

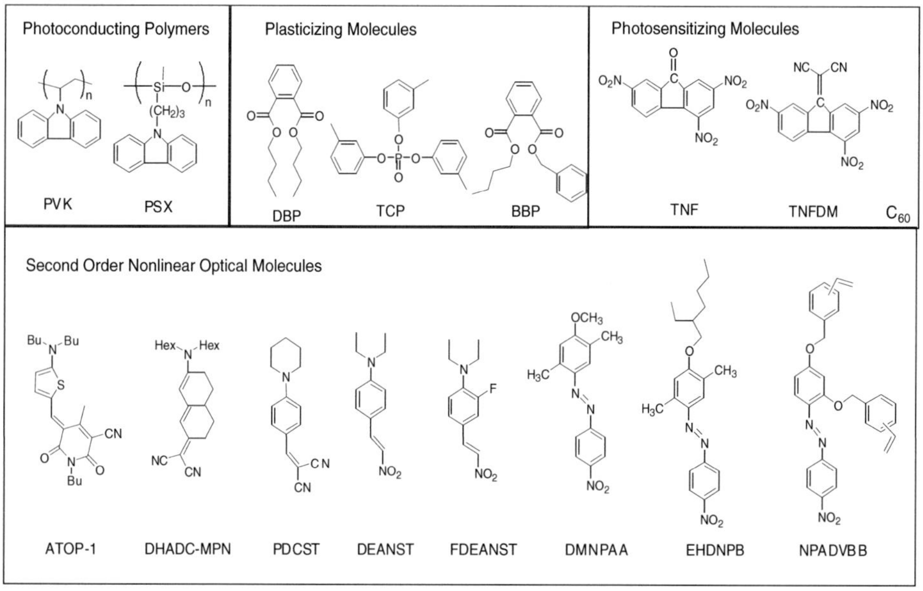

FIGURE E-3

Polymers and molecules used in preparing photorefractive composite films.

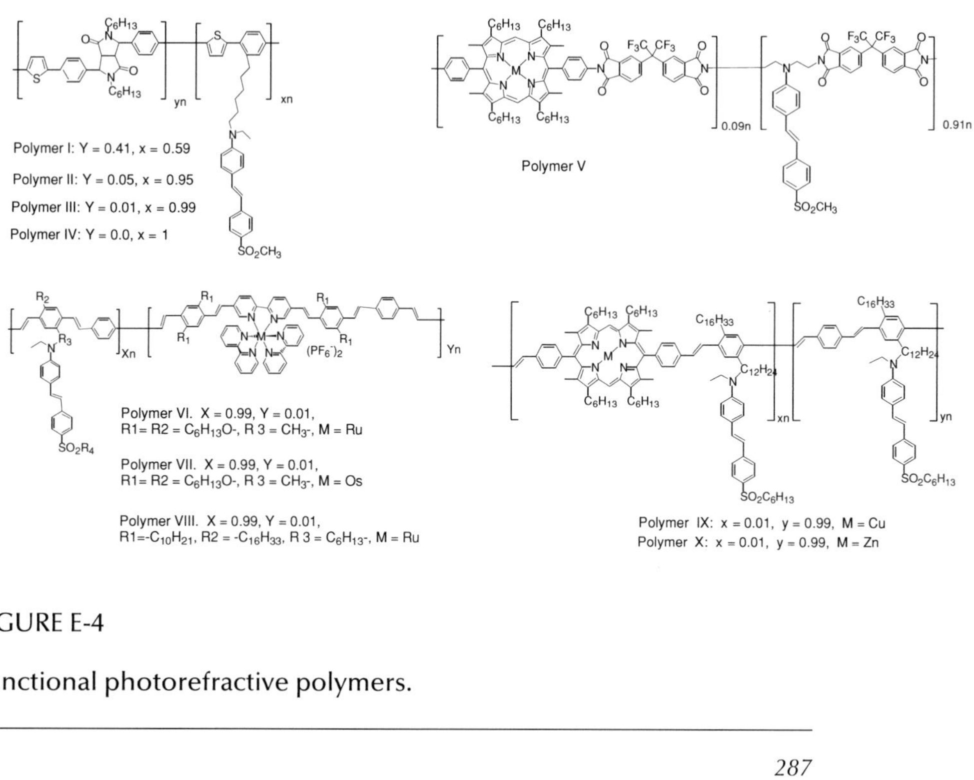

FIGURE E-4

Functional photorefractive polymers.

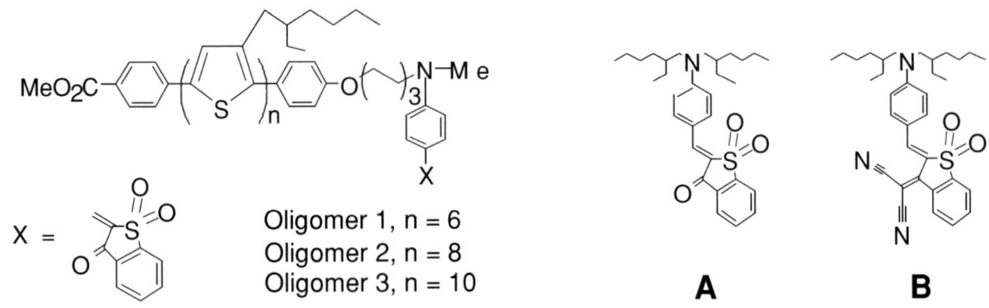

FIGURE E-5

Monolithic molecular photorefractive materials.

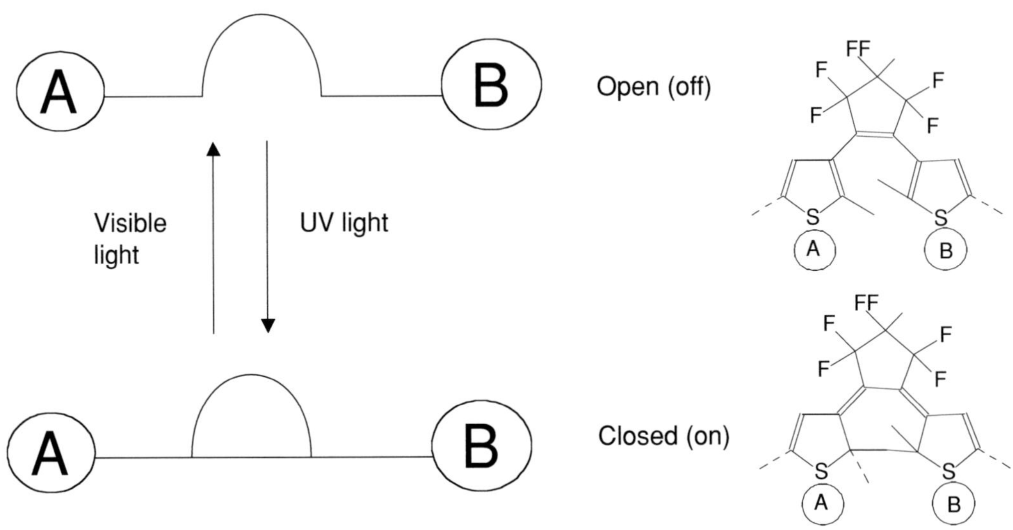

FIGURE E-6

Photochromic switch. SOURCE: Reprinted by permission from Fraysse et al. (2000). Copyright 2000 by Wiley-VCH, STM.

orbital structure that favors or reduces electron transfer depending upon its final molecular configuration (Fraysse et al., 2000).

The supermolecular structure rotaxane, shown in Figure E-7, is reported to have mechanical switching capabilities. In a rotaxane structure a macromolecular ring (analogous to a bead) encircles a linear molecular chain (thread) that is terminated by bulky stopper groups that prevent the ring from disengaging. The linear molecular chain also contains polar/nonpolar regions or docking stations that control the transport process. In the example shown in Figure E-7, the macromolecular ring consists of two positively charged bipyridinium groups separated by p-xylyl groups. The linear chain consists of flexible polyether groups segmented between benzidene and biphenol electron donating groups. Migration (switching) of the ring between the benzidine or biphenol stations depends upon temperature and oxidation reduction (Bissell et al., 1994).

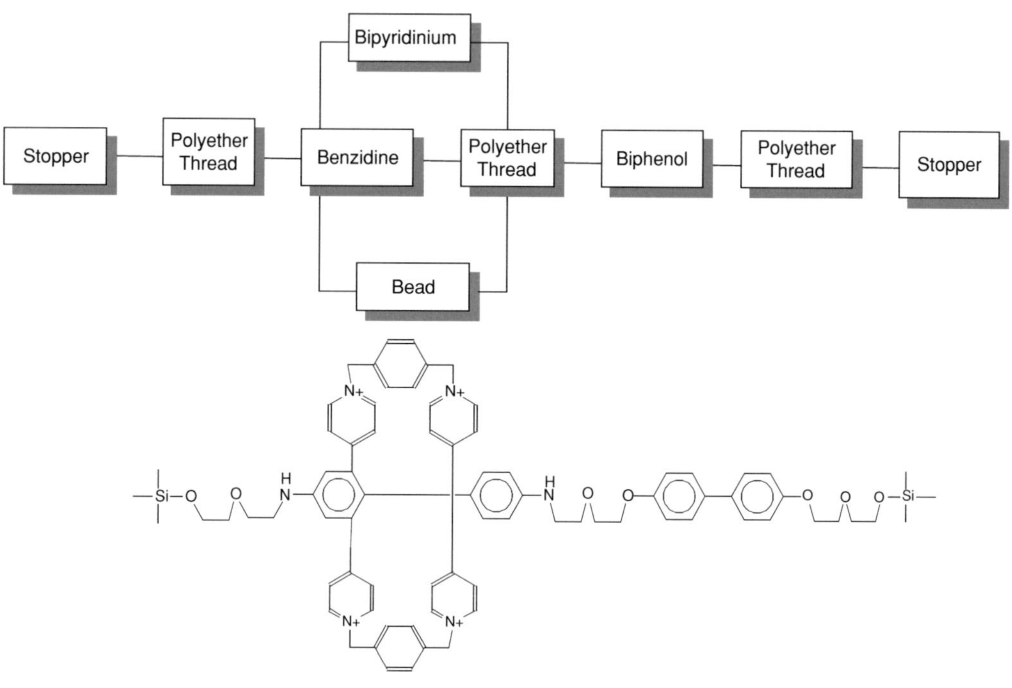

FIGURE E-7

Bead-on-a-thread molecular switch. SOURCE: Reprinted with permission from Bissell et al. (1994). Copyright 1994 by Macmillan Magazines Limited.

MULTICOMPONENT ORGANIC MATERIALS FOR ENHANCED CARRIER MOBILITY

Multicomponent organic materials can enhance carrier mobility by providing multiple absorption pathways and high interfacial areas between electron and hole transport regions, providing a mechanism for exciton dissociation. One approach is shown schematically in Figure E-8.

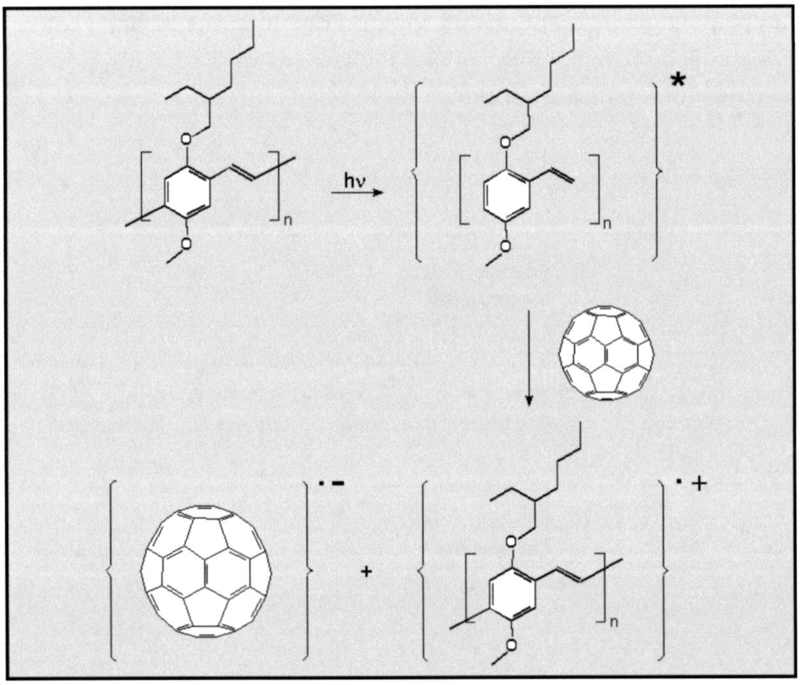

FIGURE E-8

The photo-induced electron transfer from a conjugated polymer (MEH-PPV) to C_{60}. First, the conjugated polymer absorbs a photon of light and becomes electronically excited; second, this excited molecule transfers an electron to a nearby C_{60} molecule within 3×10^{-13} seconds. The result is a radical-cation/radical-anion pair. SOURCE: Reprinted with permission from Yu et al. (1995). Copyright 1995 by the American Association for the Advancement of Science.

REFERENCES

Bissell, R.A., E. Cordova, A.E. Kaifer, and J.F. Stoddart. 1994. Chemically and electrochemically switchable molecular shuttle. Nature 369:133-137 [see also Borman, S. 1994. Reversible molecular switch synthesized. Chem. Eng. News, May 16, pp. 8-9, and Borman, S. 1999. Key step made toward molecular computing. Chem. Eng. News 77(29):11-12].

Fraysse, S., C. Coudret, and J.-P. Launay. 2000. Synthesis and properties of dinuclear complexes with a photochromic bridge: An intervalence electron transfer switching "on and off." Eur. J. Inorg. Chem. 7:1581-1590.

Yu, G., J. Gao, J.C. Hummelen, F. Wudl, and A.J. Heeger. 1995. Polymer photovoltaic cells: Enhanced efficiencies via a network of internal donor-acceptor heterojunctions. Science 270:1789.

APPENDIX F

Bioinspired and Bioderived Materials

In this detailed appendix the Panel on Bioinspired and Bioderived Materials examines what organisms can do, and it takes a flight of fancy to suggest what can be "borrowed" for advanced materials or processes. The panel writes from the perspective of what biology does, rather than what defense needs are, because it does not know what applications will arise from the use, adaptation, or mimicry of any given biological structure or design. Even where an application can be imagined with some specificity, it would be a mistake to assume it to be the only application that could arise from that biological model. Examples are

- *Atomic level control of structure—materials by design.* One of the most striking capabilities of living organisms is their ability to produce extraordinarily complex molecules with virtually error-free control of the location of each constituent atom, down even to control over the synthesis of one or another optical isomer. Molecules display sophisticated functions because each of their thousands of atoms has been placed in a precise, predetermined position in the molecule. Further, organisms design molecules they need, with the properties required, by controlling the number, type, and arrangement of their atoms. Through enzyme catalysis and templated synthesis (see below), organisms make exactly what they need with little waste, a production scheme that is clearly a model for in vitro synthesis. This control of structure at the atomic level is also a way to address the critical issue in, for example, the ceramic or metal-processing goal known as net shape manufacturing, where the product is produced in the shape required, rather than being machined after production.

- *Adaptation to the environment.* Organisms sense their environments (see below) and alter their properties to adapt to them. Shifting humans to high altitude, for instance, increases the production of the molecule 2,3-bisphosphoglycerate, which binds to the protein hemoglobin in the blood, altering its shape to decrease its affinity and thus allowing the hemoglobin to deliver more oxygen to the muscles and brain.
- *Amplification of signals.* Blood clotting, gene expression, and the activation of enzymes in the control of cellular energy production all involve the amplification, by many orders of magnitude, of signals initiated by a very small number of molecules, photons, electrons, or ions. Amplification is a multistep pathway, each step activating an enzyme that then activates a very large number of copies of the next enzyme in the pathway. Amplification is often performed without additional energy input, a feature that has great potential value.
- *Benign processing.* Biological processes are generally less hazardous than their synthetic counterparts. Synthetic nanocrystals, which are of such great interest now, are synthesized at very high temperatures with hazardous precursors (see the work of Alivisatos at University of California at Berkeley, Bawendi at Massachusetts Institute of Technology, and Mirkin at Northwestern University). Organisms produce magnetic and semiconductor nanoparticles and other materials, often with great homogeneity, at room temperature and pressure (e.g., *aquaspirillum magnetotacticum*).
- *Biocompatibility—interfaces with living and nonliving materials.* Organisms have learned to control a wide variety of interfaces with disparate materials. Biomineralized tissues of a variety of compositions are in contact with organic materials in bone, teeth, and shell. On the other hand, the use of inorganic materials for implants, while successful in the short term, often encounters long-term problems at the interface with the living tissue.

A great deal of effort is being expended to understand and mimic the structure and properties of mineralized tissues. The ability to "grow" armor (see Chapter 7) in specific shapes, much as a lobster does, would be of great benefit, as would the controlled growth of mineralized phases for functional thin films or particulate applications (Klaus et al., 1999). Phage display and other such techniques are being explored with some success (Seeman and Belcher, 2002) in the specific binding of proteins to semiconductor and other nanocrystals. The crystalline ordering of the viral particles could allow the alignment of the nanocrystals in arrays that might have applications. Both of these efforts involve the ability to control the phases and precipitation of inorganic materials through the use of designed proteins.

Attempts to make hybrid structures and devices composed of both inorganic and organic or cellular materials have also encountered difficulties, primarily in terms of long-term cell viability in such a foreign environment. This interface is becoming increasingly important because we must envision hybrid devices that exploit the many cellular functions that cannot yet be reproduced without the living cell but that are required for a particular function. For many years, whole cells have been attached to surfaces in bioreactors for fermentations. Hybrid circuits with both semiconductor chips and synaptically connected neurons have been explored. Nerve cells from the snail *Lymnaea stagnalis* have been immobilized, through nonspecific linkages, on silicon chips, using polyimides. Interfaces with the neurons were such that voltage pulses on the chip could excite the cells (Zeck and Fromherz, 2001). "Metabolic engineering" has been used to alter the surfaces of cells to improve their binding to specific sites on nonliving surfaces—metals, polymers, or ceramics—while the cells maintain their natural functions (see the work of Carolyn Bertozzi at University of California at Berkeley).

- *Biodegradable materials.* Living things, almost by definition, make biodegradable materials. Biosynthesis is accomplished almost exclusively by enzyme catalysis of chemical reactions. These types of bonds are also broken enzymatically, by enzymes widely present in the environment. A fundamental law of enzyme catalysis requires that enzymes increase the rate of the reaction in the forward (synthetic) direction to exactly the same extent as they increase the rate of the back (degradative) reaction.
- *Biopolymers: control of properties.* Biopolymer molecular backbones are constructed of a far greater variety of monomeric units than are synthetic polymers and thus can have a far greater range of properties and can have those properties tuned to a much greater extent. Monomer placement in the molecular chain is, in most cases, precisely controlled, giving rise to macromolecules of exact chemical composition and degree of polymerization. Nucleic acids use five primary monomers and many others to a lesser extent. Proteins use 20 primary amino acids, some of which are modified after initial synthesis. Although some carbohydrates are

composed of a single monomer, others draw from an alphabet of many more that vary in size, isomerization, charge, and functional group and each affect the properties of the polymer in a different way.

Recent research has progressed well in expanding the number of monomeric units in both proteins and nucleic acids, allowing the incorporation of monomers with an almost unlimited variety of sizes or with specific redox, optical, electrical, magnetic, and chemical properties. Equally important to the materials properties of these polymers is the fact that biopolymers, whether nucleic acids, proteins, or carbohydrates, are made to a precise, uniform length, giving greater control of properties and also the potential for enhanced properties, alignment, and crystallization. Naturally occurring biopolymers do have a limited number of different backbone structures, but research has broadened this range, further increasing the breadth of properties that can be achieved. In addition, techniques are being developed to use these materials in unusual environments. Mirkin, for example, has developed a technology to "write" thin lines of proteins or nucleic acids on a variety of inorganic surfaces using the tip of an atomic force microscope as a quill and a solution of the polymer as the ink.

—*Nucleic acids.* DNA and RNA are of course involved in the storage and use of cellular information, and RNA in catalysis and as scaffolds in supramolecular structures. These materials are found base-paired in the well-known "double helix" form of DNA or in single strands, which often, through the same mechanism, fold back on themselves. DNA double helixes are in fact structural materials, exhibiting the properties of long, relatively rigid rods with a persistence length of approximately 50 nm. Folded single strands, as in transfer RNA, form precisely determined three-dimensional structures. These structures are critical to the role of these molecules in genetic information storage and transfer, but they can also be exploited in other materials applications. Alivisatos, Mirkin, Seeman, Frechet, and others have linked nanocrystals or photoactive groups with electronic or optical properties to individual single strands of DNA of defined sequence and have synthesized structures through base pairing that link several single strands to align the active groups in precise positions in space. Some of these structures have demonstrated energy transfer from donor to acceptor groups, presenting intriguing possibilities for electronic structures. Ghadiri and others have developed systems that exploit the base pairing capability of DNA for rudimentary "computing" devices (Mao et al., 1999). These results are tantalizing at best, with no clear path to application, but no clear evidence exists that applications cannot be achieved.

—*Proteins*. Driven by their amino acid sequence, proteins fold into precise three-dimensional shapes. The range of properties that these polymers exhibit or contribute to is enormous, including lubricity, adhesion, viscosity, stiffness, toughness, flexibility and other mechanical properties, clarity, and, in the case of aspartame, taste.

- *Color*. We have all come to admire the beauty of the colors of living organisms, but of greater interest is the potential to mimic their use of color as a functional property. Some organisms, for example, can alter the molecular structure of their surfaces to alter their appearance as protection from predators. Certain birds (e.g., peacocks), fish, snakes, and butterflies appear colorful not (only) because they have pigments but because their overlapping carbohydrate scales impart iridescence by creating interference patterns that give the appearance of different colors depending on the angle of viewing, the wavelength of the light illuminating them, and the medium they are immersed in. The color observed can also be altered by changing the refractive index of the liquid between the layers. Use of a solvent with the refractive index of cuticle results in the complete loss of constructive interference, no iridescence, and only the dull brown color of melanin. Chameleons change in response to changes in light and temperature (not, apparently, to blend in with their surroundings) through their control of the extent to which various layers of colored or reflecting cells below their skin are exposed to the surface. Work on this control of color, which has obvious defense applications for uniforms and equipment paints, is progressing in a number of laboratories, including those of Gregory at Clemson, Hanks at Furman, and Samuels at Georgia Institute of Technology in a search for polymers that change their response to electromagnetic radiation in the presence of applied electrical or magnetic fields.
- *Catalysis/enzymes*. Biocatalysis is an exceptionally efficient manufacturing process. Biocatalysts accelerate reactions by up to 13 orders of magnitude, with exquisite specificity for starting materials and products (producing no byproducts), at room temperature and atmospheric pressure. The activity of enzymes can be controlled over several orders of magnitude through the binding of specific effector molecules. Selective activation of enzymes could effect specific chemical conversions of munitions, therapeutic agents, and odorants. Harnessing these materials would be nothing short of revolutionary.

• *Combinatorial synthesis.* Each of us carries in our blood stream more than a hundred million different antibody molecules, all with slightly different structures. As a result, in virtually all cases, there is an antibody that can bind, jigsaw-puzzle-like, to an invading virus or bacterium and inactivate it. This parallel synthesis of a huge variety of extremely similar structures can have great impact on development of materials. In many cases, the current state of theory and modeling is too primitive to allow prediction of the structure of a single optimal material with the desired mix of properties. Battery manufacturers have recently announced that their next-generation cells will contain materials discovered through combinatorial methods.

• *Computation.* The holy grail of computer designers remains a device that mimics the human brain. Nothing comes close to the brain's ability to store and retrieve information, often from apparently unrelated "data entry" events. Its ability to focus on a subset of the huge number of stimuli and inputs it receives at any one time is also unmatched, as is its ability to "reason" by combining bits of information and weighting them appropriately as it sums their input into the solution of a problem. No device competes in use of spoken language. It is true that silicon computers are already orders of magnitude faster than the brain; the brain's capabilities rely, in part, on its ability to process in parallel.

A number of biological materials are being studied for brain-like properties. The protein bacteriorhodopsin has been shown to have the capability for holographic data storage. There are initial reports of "DNA computers." Clearly, an understanding of how the brain does parallel processing and the mechanism of the self-assembly of the billions of neurons into a functioning brain will be of great benefit, with many as yet undefinable applications.

• *Conformational change.* Many of the responses to external stimuli that have been discussed in other sections of this text (adaptation, sensors, smart materials) are based on the fact that proteins can alter their structure, and therefore their properties, in response to external stimuli. The response is mediated through the binding of one or more molecules from the environment at a specific site on the protein. Enzyme activity is regulated in this fashion, with the protein shifting from an active conformation to one that is less effective in binding the substrate, performing the chemistry of the reaction, or both. The oxygen-binding protein hemoglobin also exhibits the effect. On a shorter time scale than that for the increase in the concentration of 2,3-bisphosphoglycerate, hemoglobin binds to single protons. In low-proton environments like the lungs, it assumes a conformation that enhances its ability to bind oxygen. In the more acid environment of the muscles, however, there is a higher concentration of protons. These bind hemoglobin and cause it to change shape, making it less able to bind oxygen. The oxygen it had bound in the lungs is released so it can be used in the metabolic processes that create the energy required for muscle contraction.

- *Energy conversion.* Chemical energy powers organisms without the use of flammable fuels and the volatility and heat of reaction associated with those fuels. Molecular motors, described elsewhere, convert chemical energy (in the form of ATP) directly into mechanical energy, bypassing the heat production step of conventional motors and engines. The photosynthetic sequence of light energy capture, production of "high energy" chemical stores, and the oxidation of those stores to produce metabolic and mechanical energy is one that, if harnessed, would revolutionize energy conversion. Much work is progressing on photosynthetic mimics, both biological and inorganic photovoltaics. Though many are remarkably efficient mimics of aspects of the process, much remains to be done.
- *Evolution.* All too often we encounter materials that are ideally suited to perform in the environments for which they were originally designed but poorly suited to new environments that arise later. Species evolve constantly to allow them to survive in new environments. Each population includes individuals with minor variations from the norm in virtually every property. A change in environment increases the relative reproductive success of one such variant, which then moves to dominate the population. Evolution has been recreated in controlled environments for experimental purposes. It has also been employed in the "maturation" and optimization of antibodies and other proteins, where the continual production of variants and the selection of the "better" variants has led to better products. Defense systems capable of evolving would be of obvious value.
- *Extreme environments.* In general, biological systems are regarded as impractical for many applications because of their sensitivity to extremes of environment. On the other hand, organisms have adapted to live in the below-freezing waters of Antarctica or the ocean depths and in the near-boiling conditions of hot springs or deep-ocean thermal vents. Those living at the ocean depths are protected against exceptionally high pressures, those living in salt flats are protected from high osmotic pressure, those that line the stomach are adapted to extreme acidity, and others are

exceptionally resistant to ionizing radiation. In each case, specific protective mechanisms have evolved. In principle, transferring these to organisms or systems with the required properties could protect them from these extreme environments.

- *Hierarchical construction.* Biological structures are extraordinarily complex, far more so than synthetic systems, and their sophisticated properties reflect that complexity. However, their synthesis is usually far less complex, relying on sequential hierarchical construction principles. For example, the synthesis of collagen fibers, whose thickness can be measured in millimeters, can be simply described as a sequence of construction steps, starting with the association of individual atoms and then associating the resulting structures in an increasing complex manner. The path is programmed into the structure of the material, allowing great complexity of result with minimal complexity of design.
- *Lightweight materials.* Living systems are inherently lightweight. They use almost exclusively the first row elements—carbon, hydrogen, nitrogen, oxygen—with lesser amounts of phosphorus and sulfur and very small amounts of others. Our ability to mimic living systems and create structural and functional materials from these elements would lead to enormous reductions in fuel use, and in the weight carried by the highly equipped modern soldier.
- *Lubricants.* Enormous inefficiencies, loss of function, and expense result from inadequate lubrication of the contact surfaces of moving parts. Living systems, which must solve the same problems, have evolved molecules to lubricate joints, portions of the eye, and internal organ surfaces. These usually highly charged molecules could be models for biomimetic lubrication. Hyaluronic acid, for example, which is strongly anionic is an important component of the synovial fluid that lubricates the joints.
- *Mass production.* Large-scale production of materials can often be expensive. Regulatory mechanisms in organisms allow for the production of defined levels of each molecule made, using "inducers" or "repressors" and "weak promoters" and "strong promotors." Levels of product production approaching tenths of total cell volume can be achieved. Several years ago, the genes for the natural plastic polyhydroxybutyrate were transferred into plants. Acres of farmland devoted to these transgenic plants could be inexpensively harvested and the polymer extracted. Genes for proteins are now being inserted into goats or cows in a way that leads to their secretion

into the easily collected milk, which, of course, can be "grown" for the cost of animal feed.

• *Membranes.* The cell membrane is an extraordiary multifunctional structure. It defines the selectively impermeable boundaries of cells. It helps define the surface topology of cells; houses transport systems to move materials in or out; and incorporates motors, rotors, energy-producing devices, and exquisitely sensitive, selective sensors. It creates nonpolar compartments in the midst of a fully aqueous environment. It is self-healing and self-assembling, can grow as the cell it surrounds grows, and can split into two as the cell divides. Membranes are highly flexible and can adapt their shape to a variety of structures and also to perturbations in those structures as the cells progress through the various stages of their lives or perform their myriad functions. They are also quite robust, despite the fact that the individual component molecules are not covalently linked to each other.

A great deal of effort has gone into the mimicking of the cell membrane, much of it successful. Artificial self-assembled monolayers are used in a wide variety of efforts to study self-assembly or other membrane-associated properties. Membrane mimics are made with artificial molecules, mirroring the self-assembling amphiphilic properties of membrane lipids but incorporating greater rigidity through cross-linking, or functionality through light-absorbing chromophores, inserted channels, or molecular recognition groups. Use of other molecular components allows for multilayered membranes with their own sets of properties.

• *Molecular recognition.* Individual biological molecules have the extraordinary ability to recognize specific other molecules in a sea of very similar structures. This is the basis for the functioning of the cell membrane as a biosensor, the structure and replication of DNA as a genetic or structural material (see nucleic acids), and the specific catalysis of enzymes. The mechanism of this recognition involves several factors. The first is a geometric fit similar to that between two pieces of a jigsaw puzzle. Then, along the surface between the two bound molecules are complementary binding groups: Positive charges on the face of one molecule are adjacent to negative charges on the other molecule. Other forces to enhance binding, including hydrophobic interactions and polar interactions, are also found along the interface. Complexity is added by the fact that proteins are not fixed in shape but are quite flexible; they rapidly and spontaneously interconvert among a set of possible conformations. Generally, only one conformation of a given protein binds a particular target molecule, and the binding of that molecule locks the protein into that particular configuration. This is a powerful tool in protein design and function but it does add complexity to the task of designing proteins for specific molecular recognition functions.

- *Motors, rotors, pumps, transporters, tractors, springs, ratchets, contractile proteins.* Cellular function depends on a variety of motors, rotors, and related devices that use chemical energy to move ions, macromolecules, organelles, chromosomes, and even whole cells. A newly characterized motor packs DNA into the heads of viruses using 60 pN of force (Smith et al., 2001; Cluzel et al., 2000; Stock et al., 1999; Keller and Bustamante, 2000). The enzyme ATP synthase and the various components of the electron transport chain are themselves molecular pumps and motors, embedded in the inner membrane of the mitochondrion and using metabolic energy to pump protons across the membrane. Protons flow spontaneously back across the membrane, but only through the synthase, which rotates in response to that flow of protons. This rotation has been shown by Oster (Elston et al., 1998) and others to drive the synthesis of ATP. In reverse, using ATP releases energy for the rotation of the protein and the pumping of protons out across the membrane. Another rotational motor embedded in the cell membrane drives the high-speed rotation of flagella, the tails of bacterial cells that enable them to "swim" toward nutrients and away from repellants. Linear motors like myosin and kinesin move subcellular structures by direct conversion of chemical energy to mechanical energy.

To a great extent, the nature and structure of the molecules involved in these functions are understood and some manipulation has been achieved. Techniques like single-molecule spectroscopy allow the study of individual motors and rotors. Other techniques allow investigators to anchor these structures to solid supports and to pull on them and treat them in effect like springs, relating the force applied to the extension of the molecule. Mahadevan and Matsudaira (2000) have shown that small changes in protein subunits amplified by linear arrangements in the filaments can lead to structures that store energy and then release it on demand, creating movement. The interfacing of millions of these efficient biological devices to produce "macro" levels of power is not impossible by 2020.

- *Multifunctional materials.* Another holy grail of materials research is materials that perform several functions simultaneously. Here biological systems again point the way. Skeletal plates on the arms of the brittle star provide not only structure and protection but also focus light on nerve receptors beneath, somewhat like a compound eye (Sambles, 2001). Multifunctional scales coat the wings of butterflies, supporting the aerodymnamics of the wing while assisting in temperature control of the butterfly and providing wing colors and patterns that act as a defense mechanism against predators.

- *Nanoscale synthesis and function.* The revolution in nanoscience allowing the synthesis, characterization, and use of structures of dimensions 1-100 nanometers is having a major impact on a wide variety of fields. At these dimensions, the properties of an ensemble of molecules are controlled by surface rather than bulk properties, allowing properties of a single molecular composition to be tailored over a broad range. "Small" molecules like amino acids, lipids, and sugars are of single nanometer dimensions and proteins can range from 10 to 100 nanometers in diameter, making them ideal components of these systems.
- *Self-assembly.* Perhaps the most interesting property of biological systems is their ability to assemble individual molecules into large, complex, functional structures. Membranes assemble themselves because the lowest energy state of their component nonpolar molecules is the membrane structure itself. The assembly of biological molecules is directed by site-specific chemistry, while the assembly of synthetic polymers is often determined by external forces and statistical physics. Proteins composed of multiple individual proteins self-assemble, aligning the individual subunits precisely with respect to one another to perform a function as dependent on the relationship of the individual molecules as a watch is dependent on the positioning of its gears.
- *Self-healing, repair, damage/fault resistance/tolerance.* Living organisms are capable, to varying degrees, of self-repair and healing. Simple or very young organisms can replace entire sections of their bodies; more complex or older organisms are more limited, although the human liver, for example, can regenerate itself even after much of it has been removed because of damage or the need for donation as a transplant. On a more molecular level, membranes can repair holes and proteins can refold after being denatured. DNA polymerase, which copies DNA, reviews its own work, tearing out errors and replacing them with the correct base.
- *Smart materials/sensors.* Smart materials are those that alter their structure and properties in an almost immediate response to a change in their environment. In many cases these changes are reversible, in some cases, to varying degrees depending on the extent of the change in the

environment. In some cases the response is triggered by physical changes, such as optical, electrical or mechanical effects. In other cases, changes in the levels of specific chemical substances are detected. These "analytes"are detected by proteins and carbohydrates (receptors) whose structure is defined at the level of each atom and the spatial relationship between atoms. They are embedded in the cell membrane and extend beyond its outer surface to bind the target with high affinity and specificity (see molecular recognition, above). In some cases the response to the change is functional. In other cases, it is simply a record of the change. In bacteria the enzyme glutamine synthetase monitors the level of nine independent factors in its environment and adjusts its rate of catalysis on the basis of a summation of these inputs. Entire metabolic pathways respond to single molecules, such as hormones or growth factors. Most of these responses are triggered by alterations in the shape of the receptor on binding its target. The ability to change chemical activity, color, electrical conduction, and mechanical properties in response to changes in the environment would be valuable in a variety of applications. Perhaps the most advanced smart materials at this time, however, are sensors, which translate their detection of defined targets into measurable optical, electrical, or mechanical changes. In many cases, other organisms demonstrate discrimination and sensitivity sometimes approaching the detection of single molecules or photons. In principle, individual soldiers could distinguish friend or foe from a distance when sight and hearing are limited. Chemical or biological agents could be detected before their concentrations reach toxic levels. A wide variety of mechanisms have evolved to achieve this detection, and any one could be an important application. As miniaturization progresses and micro- and nanoscale devices are developed, sensors for extremely small forces will be required. Recent work has shown that the unfolding of single strands of RNA, which involves only the breaking of a number of hydrogen bonds, can be measured using optical tweezers, allowing us to speculate that these measurement tools could be adapted as mechanical sensors. In other work, cantilevers have been shown to allow the detection of the change in energy resulting from the binding of very small numbers of molecules, again allowing speculation about new, ultrasensitive mechanical sensors. The Melanophila beetle senses IR emission of a forest fire from 50 km. Vipers, pythons, and other snakes strike objects on the basis of their detecting minute differences radiated in temperature, with discrimination levels in the thousandths of a degree centigrade. Although investigators have had difficulty adapting this and similar systems to working devices, these problems may well be resolved by 2020.

- *Structural materials.* As described in other sections of this report, spider dragline silk, synthesized primarily from carbon, hydrogen, oxygen, and nitrogen, has long been envied for its strength-to-weight ratio, which exceeds that of steel. Other structural fibers, such as collagen and keratin, could find applications if their synthesis were within reach. Research is progressing in the use of synthetic materials to mimic biological models, for example, hydrogels that can reversibly bind water to mimic the ability of collagen to absorb shock. Extensive work has been proceeding for decades to improve our understanding and ability to mimic the "hard" materials like bone, teeth, and shell, which have exceptional combinations of mechanical properties and light weight, but they still elude our grasp. Mineralized components are usually made, as are the organic components, of a small group of simple compounds like hydroxy apatite, calcium carbonate, or phosphate. They do, however, form a wide variety of structures, with a wide variety of properties, performing a wide variety of functions, because the organic phase of protein controls the crystal structure of the mineral and a variety of such structures are possible. Hydroxy apatite, for example, is the mineral in both teeth and bone, materials that have vastly different mechanical properties. Progress is being made in our understanding of these materials. For example, Morse (2001) has recently shown that the synthesis of the silica needles in a marine sponge is achieved by the catalytic polycondensation of silicon alkoxides. The catalysts are aligned in linear repeated assemblies of the enzyme in the form of a rod, allowing the deposition of the rodlike spines around them. Morse reveals new routes to the synthesis and structural control of silica and polysilsesquioxanes.
- *Systems.* On a larger scale, the performance of biological systems often exceeds that of current technology. Sharkskin, for example, exhibits better hydrodynamic behavior than polished surfaces, an effect attributed to the organization and structure of the surface. Lotus leaves are remarkable in their ability to reject dirt. Their fine surface roughness prevents tight binding of dirt particles (and even glues), which are washed away by water repelled by the waxy coating (Barthlott and Neinhuis, 1997). Structural surfaces on insect legs allow for reversible adhesion, as do the microscopic setae on the feet of geckos, which can hang upside down and walk over almost any vertical wall (Autumn et al., 2002).

• *Templated synthesis.* Much of biological synthesis occurs through enzyme-catalyzed reactions, with the great specificity of these catalysts providing high efficiency and minimization of byproducts. Equally high fidelity is achieved through templated synthesis, with the product produced through its specific match to a preexisting model. This is true of DNA and RNA synthesis, with a single strand determining the sequence of bases to be organized in the daughter strand. It also occurs in the production of of mineral phases that are directed by underlying proteins (Mann, 1988; Knight et al., 1991; Addadi and Weiner, 1985; Sarikaya, 1999). This strategy could be valuable for the synthesis of numerous types of materials for defense applications.

• *Transport systems.* Living organisms must transport a wide array of molecules and structures, both in and out of the organism or its constituent cells and to specific locations within the organism or its cells. Membranes are extraordinarily selective in allowing materials to penetrate and pass through. Some even prevent the hydrogen ion from passing. On the other hand, membranes can create systems to specifically allow the passage of molecules of choice. In many cases these "channels" open or close in response to either voltage changes or the presence of other molecules. We might imagine clothing designed to allow moisture to pass through while blocking the transport of specific toxic agents. Finally, in some systems the channels are energized, allowing the buildup of concentration gradients.

REFERENCES

Addadi, L., and S. Weiner. 1985. Interactions between acidic proteins and crystals—Stereochemical requirements in biomineralization. Proc. Natl. Acad. Sci. 82(12):4110-4114.

Barthlott, W., and C. Neinhuis. 1997. Purity of the sacred lotus, or escape from contamination in biological surfaces. Planta 202(1):1-8.

Cluzel, P., M. Surette, and S. Leibler. 2000. An ultrasensitive bacterial motor revealed by monitoring signaling proteins in single cells. Science 287(5458):1652-1655.

Elston, T., H. Wang, and G. Oster. 1998. Energy transduction in ATP synthase. Nature 391(6666):510-513.

Keller, D., and C. Bustamante. 2000. The mechanochemistry of molecular motors. Biophysical Journal 78(2):541-556.

Klaus, T., R. Joerger, E. Olsson, and C.G. Granqvist. 1999. Silver-based crystalline nanoparticles, microbially fabricated. Proc. Natl. Acad. Sci. 96(24):13611-13614.

Knight, C.A., C.C. Cheng, and A.L. Devries. 1991. Adsorption of alpha-helical antifreeze peptides on specific ice crystal-surface planes. Biophys. J. 59(2):409-418.

Mahadevan, L., and P. Matsudaira. 2000. Motility powered by supramolecular springs and ratchets. Science 288(5463):95-99.

Mann, S. 1988. Molecular recognition in biomineralization. Nature 332(6160):119-124.

Mao, C.D., W. Sun, Z. Shen, and N.C. Seeman. 1999. A nanomechanical device based on the B-Z transition of DNA. Nature 397(6715):144-146.

Morse, D.E. 2001. Biotechnology reveals new routes to synthesis and structural control of silica and polysilsesquioxanes. The Chemistry of Organic Silicon Compounds, Z. Rappoport and Y. Apeloig, eds. New York: John Wiley & Sons.

Sambles, R. 2001. Optics: Armed for light sensing. Nature 412(6849):783.

Sarikaya, M. 1999. Biomimetics: Materials fabrication through biology. Proc. Natl. Acad. Sci. 96(25):14183-14185.

Seeman, N.C., and A.M. Belcher. 2002. Emulating biology: Building nanostructures from the bottom up. Proc. Natl. Acad. Sci. 99:6451-6455.

Smith, D.E., S.J. Tans, S.B. Smith, S. Grimes, D.L. Anderson, and C. Bustamante. 2001. The bacteriophage phi 29 portal motor can package DNA against a large internal force. Nature 413(6857):748-752.

Stock, D., A.G.W. Leslie, and J.E. Walker. 1999. Molecular architecture of the rotary motor in ATP synthase. Science 286(5445):1700-1705.

Zeck, G., and P. Fromherz. 2001. Noninvasive neurielectronic interfacing with synaptically connected snail neurons immobilized on a semiconductor chip. Proc. Natl. Acad. Sci. 98(18):10457-10462.